OPPOSITES: The Law and Order of the Universe

By LaVon W. Finlinson

Volume I

Doctrine and Covenants 130:20-21

"There is a law, irrevocably decreed in heaven before the foundations of this world, upon which all blessings are predicated—and when we obtain any blessing from God, it is by obedience to that law upon which it was predicated."

Opposites: The Law and Order of the Universe

by LaVon W. Finlinson

The Religious Research Association has been involved in disseminating the scientific research contained in this book since 1984. It is now the sponsor of this book. Although the association was formed on April 29, 1982, for the purpose of identifying and informing the public of untruths being conveyed against the Church of Jesus Christ of Latter-day Saints, it was independently organized and other than its officers and directors holding individual membership in the Church, it is in no way connected to or recognized by the Church of Jesus Christ of Latter-day Saints.

Cover: Photo of Haley-Bopp comet or "The Great Comet of 1997" by Jerry Lodriguss, authorization for use through Astropix LLC.

Richard and LaVon Finlinson
1-480-834-1559
rfinlinson@cox.net

Publisher's - Cataloging in Publication
(Prepared by Likes Publishing)

Finlinson, LaVon W.,

Opposites: the law and order of the universe, vol. 1 / by LaVon W. Finlinson — Pleasant Grove, Utah:Likes Publishing, 2006

xviii, 171 p.; 26 cm.
ISBN:0–9660635–6–2
1. Science — Religious aspects. 2. Religion and science. 3. Mormon Church — Doctrines.

BL240.3
261.55—dc22 2006922485

Second Edition, Revised

Likes Publishing

Pleasant Grove, Utah

TABLE OF CONTENTS

Table of Contents v
Preface vii
Credit and Appreciation ix
Introduction xii
Tribute to My Sister – Helen Pawlowski xiv

SECTION ONE – THE LAW OF OPPOSITES

The Pattern of Creation 1
The Pawlowski Model of the Atom 9
The Uniting of Bohr's Chart and the Pawlowski Atom 25
The Building of the Pawlowski Atom 31
Visualizing the Exacting Order Within the Pawlowski Atom 43
Opposition in All Things 53

SECTION TWO – EVIDENCES OF TWO OPPOSITE REALMS

The Pawlowski Model Clarifies Religious Truths 63
Two Opposite Forces – Two Opposite Ways .65
The Scriptures Testify of the Law of Opposites 67
Two Creations: Heaven and Earth – Spiritual and Physical 69
Why Did God Create the Heavens and the Earth? 71
The Garden of Eden: The Beginning of Opposite Choices 75
Mankind Enters the World to Be Tested 77
Communicating with Heavenly Father 81
The Universe is Governed By Law 83
Why Do We Associate Satan with the Negative? 89
Will We Resist Satan's Damning Ways? 95

SECTION THREE – THE LAW OF WITNESSES

Jesus Christ: God's Greatest Witness .... .99
We Must Be Found on the Right Hand of the Father 103
Complementary Opposites: The Male and the Female 109
The Word of Wisdom: A Principle Founded Upon Opposites 111
Stay to the Right, Like the Fig, and Live! 119
Atonement/Judgment: Mercy Offered, Yet the Law Sustained 129
We Must Repent, But Little Children Need No Repentance 135
We Must Be Born Again 139
The Root and the Branch – The Giver and the Receiver 143
Faith Without Works is Dead 147
Eternal Progression is Like the Positive Moving Forward 151
The Son of God 155

SECTION FOUR – ONENESS COMES FROM OBEDIENCE TO LAW

Will It Be Balance in All Things-or-Koyaanisqntsi? 159
Only By Turning to God Can We Cure Our Ills 163
God Exists and He is at the Helm 167
All Truths to Be Gathered into One 169

PREFACE

When gazing into the night sky, have you wondered, even as I, why there are so many stars in the heavens and for what purpose were they created? The immensity of space fills us with awe and reverence, yet, something deep within us whispers that we are important in the great scheme of things and that there is purpose to our earthly existence.

To my dear friends, not of the Church of Jesus Christ of Latter-day Saints, it is my hope that by presenting this material, it might engender within you a desire to know more about your beginnings. *Just who are you? Where did you come from? Why are you here? And where are you going?* Is life nothing more than the daily ritual of rising with the sun to play out life's constant challenges, then retiring as the earth continues on its never ending journey around the sun? Surely there is more to life than this. The answers to these perplexing questions have eluded most of mankind since the dawn of history and continue to puzzle the world.

Perhaps they could be answered to your satisfaction if we were to look at the world in a different way. This is why I have written this book. It is to give a new perspective to what we already know. But before trying to answer these age old questions, let us first consider the following questions: What about the elements from which we are made? Shouldn't they be a factor in the equation of life? Surely they are important to the overall picture, since all things are made up of elements. But do we ever hear anyone talk about elements when pondering the meaning of life? Of course not, but to me they are the missing pieces of the puzzle. Once I realized that a law of opposites was governing the elements of the universe and that both science and religion were based upon this law, a more cohesive order was brought to my mind and my horizons were expanded in a most dramatic way. I began to look at things in an entirely new light. I then understood why nothing can stand alone, for just as the industrial world needs positive and negative energies to create power, magnetism and light, mankind needs these same positive and negative energies to create and preserve life on this planet Earth.

Opposites start with small things then expand into large things. First we have the atom, with its positive and negative particles of energy creating energy and light. Then in the plant kingdom, the energy of the sun and the elements of the earth combine to create living things. As long as these two energies remain balanced, life is sustained. Even the heavens, with its numberless stars and planets dotting the sky, must have opposite energy forces to exist.

Understanding elements and opposites is the key to understanding the physical world around us, but by turning this key, the unseen spiritual world is opened to our view. Opposites are the very foundation of our existence, for look what happens when our spirit body leaves our physical body. We no longer exist as mortal beings. The need for opposites cannot be overemphasized, for without them there would be no life.

God has devised a plan for us mortals that takes into account the opposite nature of things. He knows that since we have both a spiritual and a physical side, we have two different natures with needs that must be met. We daily feed our physical side; indeed if we were to neglect this necessity, we would soon die from lack of life-sustaining nutrients. But what about the spirit side we cannot see? Should it not also be strengthened and nourished so as not to die from lack of attention? The answer is yes, most certainly, for if we were to neglect this unseen part, we would suffer in ways difficult to understand. Perhaps this is why there is so much confusion. With wars and contentions continually being spawned throughout the world, we are left to wonder why we are so out of harmony with one another. Could it be we are spiritually malnourished and cannot recognize the spirit of truth? Or are we listening to the wrong spirit? It seems that both are part of the problem.

This book will take a closer look at how opposites profoundly affect us and will give new insights on the following questions:

1. Who am I? - Why am I here? - Where did I come from? - Where am I going?
2. Is there a Supreme Being guiding the affairs of men?
3. Are we children of this Great Being?
4. Did Jesus Christ create the world?
5. Is Satan real?
6. Why is the physical body important?
7. Why must we become whole before entering the kingdom of God?
8. Why have prophets said we must be found on the right hand of the Father?
9. Will science someday testify of religion?

The unrest we see in the world today gives evidence that most do not have the answers to these probing questions, because once satisfied, one is better able to handle the trials and sorrows that come. Not that the pain is completely removed, but a quiet peace which "surpasseth all understanding" distills upon the soul the knowledge that all will be well. Life's adventure then holds meaning and purpose. This is why it is important to know that if we listen to God's plan and implement it in our lives, our journey on this planet Earth will be successful. But it is equally important to know that if we do not, we will suffer unnecessarily. It is my fondest hope that this book will help in your quest for truth; that you might know the true purpose of your sojourn here on earth.

CREDITS AND APPRECIATION

The science contained in this book is based on new knowledge that shows how the atom is structured and how it behaves. Science has provided some verbal descriptions of the atom, but never an exacting visual picture. It now has one.

In 1984, my sister Helen Pawlowski, of Orem, Utah, introduced me to this new thinking by showing me her Circular Periodic Table/Model of the Atom. It was obvious that the exacting order it brought to the elements was a major step forward in the field of science. These new insights inspired me to write *Opposites: The Law and Order of the Universe*. Since the original ideas were not mine, much of the credit must be given to Helen, since she provided the foundation upon which I could build and expand. Her model and many of the ideas came from a series of letters she sent to me in 1983. These, as well as her unpublished book *Mankind, A Compound In One*, have been my inspiration. To her, I express my love and admiration for persevering in studying a subject most would find unexciting. Because of her insatiable curiosity and tenacity, she has found what many think is the connecting link between science and religion. I will be forever grateful for her sacrifices in bringing this knowledge to light.

It was through Helen that the model of the atom came to be, but the ultimate source of inspiration came from our Heavenly Father and His Testator of Truth, the Holy Ghost.

I would like to express appreciation to my husband Richard for his unwavering love and support during the many long years that it has taken to write this book. His quiet, steady spirit nourished my soul. Without his confidence and sacrifice, I certainly could not have continued on such a strenuous course.

I appreciate the tremendous work that my brother Paul Williams has done over the years. He has sacrificed time, effort, and money to prove that Helen's theory is correct. He has extended her model of the atom into the laws of physics, thereby unlocking the secrets of the unseen world and those of outer space. His efforts have been significant and will prove to be of great worth.

I am especially grateful for the love, support, and confidence that Leonard and Carole Richards have given me during the time it has taken to complete this work. They are faithful and true friends who have stood by during the good times as well as the bad. In his position as president of the Religious Research Association and hers as secretary, they have been key players in bringing this work to fruition. When a challenging situation arose, they were always there willing to help solve the problem.

To my dear friend Kathryn Paulsen, I express thanks for her friendship, support and radiating enthusiasm. She has been my teacher, confidant, and friend. Her insights were invaluable during the early years in helping me better understand the scientific parts in which I was not so familiar. She has fearlessly defended Helen's theory, never hesitating to move it forward.

Sharon Swanson and Tamara Thomas played significant roles in clarifying my thoughts and ideas during the final editing. With gentle persuasion, they gave me the courage to change parts that needed changing and to move sections that needed to be moved. I cannot thank them enough for their efforts in smoothing out the rough parts that I could no longer see.

Thanks to Larry and Darla Walker, Jackie Mallyn, Sarah Hinze, Robert Zelewski, Jim Baker, Morgan Tanner, Trish Dahle, and Jay Merrill for their insights, suggestions and corrections along the way.

I give special thanks to Phyllis Likes of Likes Publishing in Orem, Utah, for refining my illustrations and professionally setting up my book for publication. She, too, helped coordinate my thoughts and gave many valuable suggestions. Phyllis is a most gracious lady that was Heaven sent to help me at the right time.

I appreciate James Tanner and Dennis Lambson for time they spent in doing the legal and accounting work for the Religious Research Association. Their efforts were timely. Also, to James Tanner's daughters, Amy, Tiffany, and Eliza, for organizing my manuscript when my computer skills were nil.

I also appreciate the interest and support that my family, parents, and friends gave me during the writing of this book, including my children Allen, David, and Diane, for helping me solve my computer problems and calming my fears.

***

In writing this book, I have used scriptures from the Bible, Book of Mormon, Doctrine and Covenants, and Pearl of Great Price so that you might know how God continues to give truth to men in all ages through His chosen prophets. By interspersing them throughout the book, you will see how they blend together and complement one another, as all witnesses of truth should. They have been brought forth by the power of God to testify of Jesus Christ and his mission here on earth. They also testify that these two great beings were the creators of both the world and the universe in which we live!

Following is a short overview of what each book contains:

Bible—Book of Mormon: These two books are a collection of the writings of various individual prophets, each testifying of God's dealings with His children during their particular time. The Bible is a testimony of God's dealings with man on the Eastern Hemisphere, whereas the Book of Mormon is a testimony of His dealings on the Western Hemisphere. Their main purpose is to testify that Jesus is the Christ, the Savior and Redeemer of the world and that only through him are we saved from the negative consequences of this earthly estate.

Doctrine and Covenants: This book is divided into 138 sections. Each section contains revelations from the Lord. They were given mainly to the prophet Joseph Smith for the purpose of instructing him on how to restore the Savior's ancient church. Since the restoration was a "line upon line, precept upon precept" type of instruction, the revelations in one section build and expand on the one before it.

Pearl of Great Price: This book contains two books of ancient origin: The book of Moses and the book of Abraham. Each clarifies the nature of God and His creations. The truths they reveal expand and enlighten the mind in a way that only new knowledge from heaven is able to do. The book of Moses reveals additional information on Satan's rebellion in the pre-existence and how it has affected the world.

LaVon W. Finlinson
Copyright, September 2004

INTRODUCTION

This work is about law: the law that governs the universe and the law that governs mankind. This law is called the law of opposites or it could be said, the law of balance and equilibrium. It is the governing law that sets the boundaries in which we must operate. Where is this law found? It is within the atom. It is here that matter is organized into two separate energy fields, thereby making it possible for power, energy, magnetism, and light to be generated and brought into a balanced whole. Because this law is embodied within all the elements, it governs both science and religion. Even math is founded upon this law.

How was this law found? When Helen Pawlowski corrected the Standard Periodic Table of Elements, this governing law was opened to her view. The structure and behavior of the atoms were revealed once the electrons were placed in exacting positions in a circular configuration. The periodic table then became a model of all the atoms.

She was able to do what seemed impossible by following Bohr's Electron Distribution Chart with exactness. This chart was the beginning of quantum, meaning the math part of science. Years ago, her interest was piqued when she noticed a 2-8-8 pattern being manifest throughout the chart. This interest was heightened when she found areas of shifting electrons that science could not explain. She felt that perhaps these areas of uncertainty were the cause of the probability/chance mentality that science had embraced. She set out to find out. As the work progressed and evidence supporting her work began to mount, it became evident she was discovering the Grand Unifying Theory science was looking for, as well as a type of "Rosetta stone" that could unlock the code of the scientific world. These are bold words, but truth calls for boldness.

Three books have been written from Helen's findings: *The Visualization of the Atom*, which is her original chemistry research; *Evidences and Ramifications of the Pawlowski Table/Model of the Atom*, which is our brother, Paul Williams's, extended physics research; and *Opposites: The Law and Order of The Universe*, which is my simplified version of the two.

As time drew near for the *Visualization of the Atom* to be published, we had hoped for some acceptance from the scientific community to shore up our base. Our expectations were centered on the two hundred copies of the *Visualization of the Atom* that we sent on 1 August 1987 to the heads of physics departments in various universities and laboratories throughout the world. Our purpose in doing so was to establish the scientific basis upon which this work rests. The responses varied from those showing a lack of interest in the concept, to others who said "superior work," "I am impressed by the tremendous work you have sent," "I think you should consider publication in the *American Journal of Physics*," and "I can find no fault with the work." But those taking the time to respond were too few in number to feel that there had been any appreciable acceptance of the work. So we had to ask ourselves, "Can we expect the scientific community at first glance to recognize the Pawlowski Model of the Atom as the pattern of creation, the pattern of opposites upon which all the elements build, when it takes time for such a paradigm shift to occur?"

We took some comfort in remembering the work of Pierre and Marie Curie, the discoverers of radium. Even though they knew it existed, it wasn't until after stirring pitchblende four long years in an open kettle outside their dilapidated Paris laboratory, were they finally able to obtain enough to prove the fact. Just as the Curies knew radium was there and had to work to produce the physical evidence, we feel the pattern shown in the Pawlowski Atom is true and that science will yet testify of the fact. At present, the testimonies of those knowing it to be true come through the strong witness of the Spirit. But the physical witness is also there. Years of concerted effort have produced convincing evidence that all aspects of life are being governed by an opposition factor set up within the elements. Even as the Curies came to know that much hard work was required for their work to be accepted, we too, have no illusions that this work will be otherwise.

Through this book, we hope to convince the reader that the "law of opposites" is indeed the Grand Unifying Theory that science is looking for, and as such, it both governs and unifies science and religion. Helen Pawlowski, the discoverer of this knowledge, is not at this time highly esteemed. She has no position of influence and holds no degrees in the field of science, but rather has one high school chemistry class to her credit. Her understanding has come through years of individual study, logic, and the inspiration of the Spirit, rather than by the learning of man. But surely this is the way God works in His efforts to confound the wise and to bring about His purposes:

> "Now ye may suppose that this is foolishness in me; but behold I say unto you, that by small and simple things are great things brought to pass; and small means in many instances doth confound the wise.
>
> "And the Lord God doth work by means to bring about his great and eternal purposes; and by very small means the Lord doth confound the wise and bringeth about the salvation of many souls." (Book of Mormon, Alma 37: 6–7)

And so the pattern of the atom is small and is brought to light by those the world would view as simple, but I testify that these things are true and will yet confound the wise. With this as my witness, I present for your consideration: *Opposites: The Law and Order of the Universe*.

A TRIBUTE TO MY SISTER, HELEN PAWLOWSKI – 1930–2001

Helen Pawlowski passed away on 2 July 2001 of lymphatic cancer. Her sudden passing was a surprise to many and came just when her model was receiving some acceptance. Although no one among the many scientifically oriented individuals who had studied her work throughout the years had come forward to oppose it, we knew we needed their help for it to succeed. How ironic that it would come so soon after her death.

Helen was a brilliant woman who saw things in a different way. I suppose this came from having a mind that was always curious and wanting to know why things behave as they do. Because she spent her life studying the many facets of the physical world; biology, chemistry, and physics were no strangers. They were just part of an ongoing process of connecting every aspect of the physical world into a oneness of purpose. But her perspective was not just from science, for her faith in an Omnipotent Creator always came first. Because she came to believe that the elements of the spiritual and physical worlds were inextricably bound together by a law of opposites, it was not hard for her to *visualize* this in her mind. But not until the winter of 1983–84, was she able to *verbalize* this thinking. It was during the development of the Pawlowski Model of the Atom that her thoughts and ideas of the previous thirty-five years came together to form her book the *Visualization of the Atom*. This gave her the needed evidence to show how opposites were being set up within the elements and how heaven and earth were working together as one.

I believe Helen's work will dramatically change our perspective about law. Since many think law is not a concrete thing, but rather a man-made device that keeps things in order, her model of the atom presents a new and exciting picture. It gives us a visual pattern of how the law is embodied within all the elements throughout the universe. Because of this universal order, we cannot escape it, nor act as if it were not there, since it is embodied within the elements of our own bodies as well! It also gives evidence that the laws of heaven and earth do not contradict one another, but rather fit together like a lock and key. Her exceptional work presents another perspective on why prophets and apostles throughout the ages have given us commandments. With two different energy fields at work within our bodies, the prophets' inspired words have informed us on how balance and equilibrium can be established between the two.

It was difficult for me at first to understand the intricate details of Helen's thinking, for my mind was not scientifically oriented, likes hers. However, I knew from the moment I heard the concept of opposition, that it was true. As I gradually picked up a truth here and there, I soon found I could verbalize it only from my layman's perspective. So this is how I have written my book. While it represents a simplified version of her thinking, Helen certainly would not have explained her work in the same way I have. However, since she was my teacher and her Pawlowski Model of the Atom the foundation of my thinking, in my mind she is a co-author. And so I pay tribute to a great woman whose only desire was to promote truth in people's minds. My desire is that I might disseminate these truths that were so dear to her heart.

LaVon H. Finlinson

– SECTION ONE – THE LAW OF OPPOSITES –

The Pattern of Creation

One evening while watching television, I happened on to a very unusual program. It caught my attention because I had never seen one quite like it. For well over an hour not a word was spoken. There was just unusual music with unusual pictures moving across the screen in a panorama of powerful scenes. What was this visual message that needed no words? It was *Koyaanisqnts*i (ka.yaa.nis.qatsi), a Hopi Indian word meaning crazy life, life in turmoil, life out of balance, life disintegrating, and a state of life that calls for another. It vividly portrayed man's frantic search for happiness, of his running to and fro with no apparent purpose to his life and of confusion that never seemed to end. The futility of man's existence was of course exaggerated, but even then, I had to ask myself, "What is wrong with the world and why are things in such chaos? Could it be that mankind is out of balance?"

I think most would agree that man's life is out of balance, in turmoil, and disintegrating. As we see wars, confusion, and unrest engulfing the world, it is obvious something is wrong. Man is so at odds with himself and all that surrounds him, one is left to wonder, "What is causing this unrest and what is to be done?"

Let us consider: Could there be laws within man's body that are being broken? If so, could a lack of understanding these laws be part of his confusion? Could the secrets of his creation be locked within his elements? I think the answers are yes, for new evidence strongly supports the premise that man will find the answers to his unrest, not only within holy writ, but within his elements as well.

But you might ask, "What does confusion or unrest have to do with elements? Aren't they just the materials that things are made of?" To better understand the close relationship we have to the elements and all that surrounds us, a few questions need to be asked:

What is the universe made of?
What is the world made of?
What are plants and animals made of?
What is man made of?

The answer to all these questions is *elements*! All creation is made up of the same 92 basic elements in different combinations and in differing levels of complexity. These could be put into two categories: the positive and the negative. You would think with such a diversity of creations, more would be needed. But no, it just takes two: positive elements that give and share electrons and negative elements that receive electrons. With such opposite attributes, they are attracted to each other and can fill one another's needs. So how did the two come together when the above things were formed? Was it by chance or could it have been by exacting law?

Evolution or Exacting Law?

As the marvels of the scientific age unfold before our eyes, we might wonder why scientists are now able to accomplish such amazing feats. Is it because they have come to better understand the laws that are embodied within the elements? Most certainly, for they know that in order to utilize the elements, they must know the laws that govern them, and that success is assured if these laws are obeyed. Because of their confidence in law, astronauts have gone to the moon and back; numerous planets have revealed their hidden truths and space is being explored in ways previous generations never thought possible.

With these realities in mind, a question needs to be asked, "If the elements used in the marvelous scientific advancements of our day have laws within them which govern their behavior, does it not seem likely there are laws within our bodies which determine what we, as human beings, can or cannot do?" Reason alone tells me this must be so, since we are made of these same basic elements. This same reasoning tells me that if we apply these laws in our lives, just as science has applied them to their endeavors, it will make the difference between our having a successful or unsuccessful journey on this planet earth.

Why then, is there unrest and confusion in the world? Humans are made from these same elements, yet we continue to flounder from a seeming lack of direction. Could it be that we do not know the importance of these laws? If we did, we would surely want to obey. But how are we to know them? Scientists of great learning have claimed that we evolved from a simple one-cell protein and that it took billions of years to bring us to our present state. But did life really start on such a tenuous thread that its very foundation was nothing more than a lightning bolt activating proteins in the ancient primordial sea? According to the big bang and evolution theories, man came into being with no plan or blueprint, no previous model to work from, no thoughts or ideas involved. It was much like a house that built itself, with no ability or plan needed to bring about such a complex creation. These theories, however, leave much to be desired, for this "by chance" sort of creation sounds so effortless—as if we can do nothing but wait for our destiny to unfold and that success and failure are not measurable since there are no set laws by which we can be measured.

Recently, theories concerning evolution have increasingly come under attack because of the many unanswered questions they have spawned. One would think if they were all true, they would have brought peace and contentment to our souls. Instead, we hear the constant cry: "Who am I, where did I come from, and why am I here? How did this vast universe come to be?" These evolution theories have not and will not succeed because they do not reveal the purpose of our existence.

The following strongly suggests that current theories leave much to be desired in man's search for his beginnings. It quotes Mary Leakey, the noted archaeologist whose discovery of ancient skeletal remains in Africa brought about a revision of the history of man, as saying, "*The fact is we really know very little about man's origins*." She then goes on to describe her field of labor. She says there are two different classes of archaeologists: the "armchair archaeologists" and the "field workers." The one teaches while the other goes into the field to dig for evidence. She then explains the weaknesses of each:

> The armchair archaeologists, mull over what others have discovered to try to think up ideas on how to interpret new finds, often without any reference to the conditions under which the finds were made. Then you have people who go out and dig things out. They may not put the right interpretation on everything but they do know all the circumstances surrounding the finds. I think they are in a stronger position than the armchair archaeologists whose speculations are fed to students not in a position to be critical...*Everyone wants to say they have found the oldest and the most significant bone. That is at the root of half the trouble.* In the end, the arguments will never be resolved unless there's more evidence. (*U.S. News and World Report*, "We Really Know Very Little About Man's Origins," Dec. 24, 1984, emphasis added)

Currently we are taught that man evolved through a process of natural selection and that the ape of the African plains was the final step before reaching our present state. How can this be since the critical evidence linking the two has never been found? Yet schools and universities continue teaching it as if it were fact. Is this the way our children will arrive at truth? It makes me think of the man who started on a journey but took off on the wrong road. No matter how far he travels or how long it takes, he will never reach his destination, for he is not even on the road that will take him there in the first place.

These next words should give us hope that in the end truth will prevail:

> Ever since the famous Scopes "monkey" trial of 1925, at which Clarence Darrow and H.L. Mencken rained ridicule on Biblical accounts of creation, the theory of evolution has enjoyed the status of revealed truth for most of the educated public. *But there are cracks in the facade.* The great body of work deriving from Charles Darwin's revolutionary 1859 book, *On the Origin of the Species,* is under increasing attack—and not from just the creationists. Richard Lewontin of Harvard, a biologist of impeccable secular views, accuses Darwinists of telling "Just So" stories when they try to show how natural selection explains such novelties as long-necked giraffes. The editors of a new book assert that when it comes to accounting for life on earth, natural selection should be "relegated here to the (explanation of) last resort." (*Newsweek*, "Science Contra Darwin," April 8,1985, pg. 80, emphasis added)

We should seriously consider that science, in and of itself, will never be able to answer the enigma of creation nor satisfy our deep longings from within. Science does not have the full picture because it does not consider the purpose of life here on earth. Sir John Eccles, a Nobel laureate and a pioneer in brain research, found this to be true:

> *We need to discredit the belief held by many scientists that science will ultimately deliver the final truth about everything.* Science has gone too far in breaking down

> man's belief in his spiritual greatness and has given him the belief that he is merely an insignificant animal who has arisen by chance and necessity on an insignificant planet lost in the great cosmic immensity. (*U.S. News and World Report*, "Science Can't Explain 'Who Am I? Why Am I Here?'" Dec. 10, 1984, pg. 80, emphasis added)

Because we are erroneously taught that we evolved from small microbes in primordial ooze to our present state, many individuals are blinded to the greatness within themselves. Instead of seeing the spark of divinity that lies within every man, they look beyond the mark. They discount the possibility that a Supreme Being, with infinite abilities, is overseeing this vast universe and that He brings order out of chaos by organizing the elements in a specific way. We must never forget that our spaceships, computers, and satellites did not evolve into complex modern devices by themselves. They had an outside energy source (man) bringing purpose and order to that which otherwise would have remained non-productive elements. Since the world in which we live is made of these same elements, why do some claim the elements making up the world came together by chance, while the wonders of our age were brought about by man's creative genius? *They are trying to explain creation by using two sets of rules for the same elements.* We give man credit when he creates, so why not give the Creator of all things the same recognition when He creates?

I believe that in the end, science will be religion's strongest ally. Why do I think something so foreign could ever happen? It is because God is omnipotent and oversees all things. Since the world in which we live is His world, and the laws governing it are under His jurisdiction, He can change our paradigm thinking simply by enlarging our knowledge of its physical laws. I propose that this enlarged knowledge will then reveal the universes—and man's—divine Creator. Though some may dismiss His existence, it is evident to me that He is the one setting the pace of man's history as well as defining the rules of the game.

Where are the Laws Which Govern the Heavens, the World, and Man?

Rick Gore, a well-known physicist, has given us a direction on where to start looking for the laws that govern the immensity of space. He states that, "*In the structure and behavior of the very smallest things, lies the explanation of all matter.*" (*National Geographic*, "The Once and Future Universe," June 1983, pg. 710)

Why would the secrets of the universe be found in the smallest things? This is where matter is organized and the structure that defines the perimeters in which all things must operate is set up. When a pattern has been established in the smallest things, all things are beholden to work within its boundaries, even though the process of development becomes more complex and interacting. *What is the smallest thing? It is the atom.* An atom is the smallest unit of an element. Therefore, in the structure and behavior of the atom, lies the explanation of all matter.

The most brilliant scientific scholars are asking, "What is this illusive pattern that controls the elements?" Those in the field of physics call it the "Unified Field Theory."

They want to know what is keeping things synchronized and working so well together. They are looking for a simple, yet aesthetically beautiful, underlying pattern that would mathematically express every process in nature. The following quote tells of this search:

> When Albert Einstein equated energy with matter in the most famous mathematical statement of the 20th century, E=mc2, he dreamed of the day when an equation of similar simplicity and elegance could be devised to describe all the workings of the universe. Einstein never achieved that goal, but today physicists are engaged in a worldwide search for the answer to one of the hottest and most fascinating subjects in science—a Unified Field Theory—a universal formula mathematically expressed, that would account, in the words of Timothy Ferris, "for every fundamental process in nature, from the jostling of atoms to the wheeling of the galaxy. *To know the atom, he says, it seems, we must know the universe; and to know the universe, we must know the atom.*" (*PBS Undercover Magazine*, November 1985, pg. 6, emphasis added)

Many scientists feel that a few simple laws are governing matter and that even on the more structured molecular level, as simple things become more complex; a consistent pattern is governing its movements. Ceylon-born Cyril Ponnamperuma, in an American Chemical Society meeting addressed this idea. He told how extra-terrestrial life is more likely to be chemically similar to life on this planet because creation is following a universal formula. He offered what he called new clues that the genetic code, which dictates how all-living things, look and behave, did not develop randomly. Rather, he said:

> The recipe is prescribed by precise interactions of chemicals that would be identical wherever they took place...*There is a natural tendency, like water running downhill, for the genetic code to spell out words the same way every time*...It's demanded by the chemistry of these compounds...This is what is used to write the entire book of life. (*Arizona Republic*, "All Life Forms May Resemble Those of Earth," Sept. 10, 1986, emphasis added)

What is Happening Inside the Atom?

It is evident from the above statements that it is important we know more about the atom. This would then give us a better understanding of the world around us and provide a glimpse of who we really are, for what is happening inside the atom reflects what is happening throughout the universe.

First, what is an atom? An atom is made up of protons that have a positive (+) charge, electrons that have a negative (–) charge and neutrons that are neutral. Together they make up a unit of matter/energy. Although much is known about the atom, there is also a void in our understanding. How were the positive and negative elements formed? Was it by probability/chance or exacting law? Do all the elements exist as separate entities, yet are structured to work together as one entity? Do positive and negative energies unite in a random, swirling probability cloud or is there an overriding pattern that connects the two?

Perhaps this lack of understanding the structure and behavior of the atom is why evolution and probability/chance theories have ruled man's thinking. I'm sure these theories were not purposely designed to mislead, but it is easy to speculate when there is a void. Obviously, if a universal formula exists for writing the entire book of life, we need to know it to be on solid ground.

I believe this universal formula or pattern has been found! Although this is a bold statement, it is not being said lightly, for I well understand the magnitude of such a statement. Since this work has been 22 years in the making, it is now time to present it, as well as defend what has been found. In a most remarkable way, this universal pattern does show how positive and negative elements were formed and how they come together in an interacting, working relationship to produce energy, magnetism and light. It gives a visual picture of how the atoms making up the elements are able to achieve balance and equilibrium, even though the elements making up the atoms are unevenly divided with 71 positive, 15 negative, and 6 inert gases. It also shows how (+) and (–) sub particles become either positive elements with a plus (+) charge or negative elements with a minus (–) charge. It reveals how matter works together as one.

God's power is so overwhelming and breathtaking in its scope and magnitude that we can barely comprehend it. Even so, I believe this pattern gives compelling evidence of His divine workmanship. I also believe that once its dynamics are understood, we will then know He does not operate in a probability/chance manner, but is following accurate, well-defined laws. It is simple, yet complex—understandable, yet complicated. It is obvious that if He can create such a pattern, He can both organize the elements into worlds without number and into small entities that have life. Perhaps its greatest blessing will come from our expanded ability to better comprehend His works. Though we may sometimes feel small and insignificant to all that surrounds us, this universal pattern gives compelling evidence that we, as one of His creations, hold a unique and vital role in His immense, energy driven universe that vibrates, like the atom, with power, magnetism and light!

If we are to know more about our beginnings and to understand why things behave as they do, we must come to know and comprehend the structure and dynamics of the atom. It sets the pattern because it contains the law. For us to find peace and purpose to our existence, we must come to understand why these laws are important in our lives. Then once we understand them, we must learn to obey them. Should we decide otherwise, "Koyaanisqntsi" will abound and the suffering and consequences of broken law will continue to disrupt and weaken society. If only we will open our eyes to see, indeed all creation is pointing to the same conclusion: law is governing man and the universe, because the law is within the elements! The Pawlowski Atom, which I believe is this illusive pattern of creation, will now be presented for your consideration.

THE PAWLOWSKI MODEL OF THE ATOM

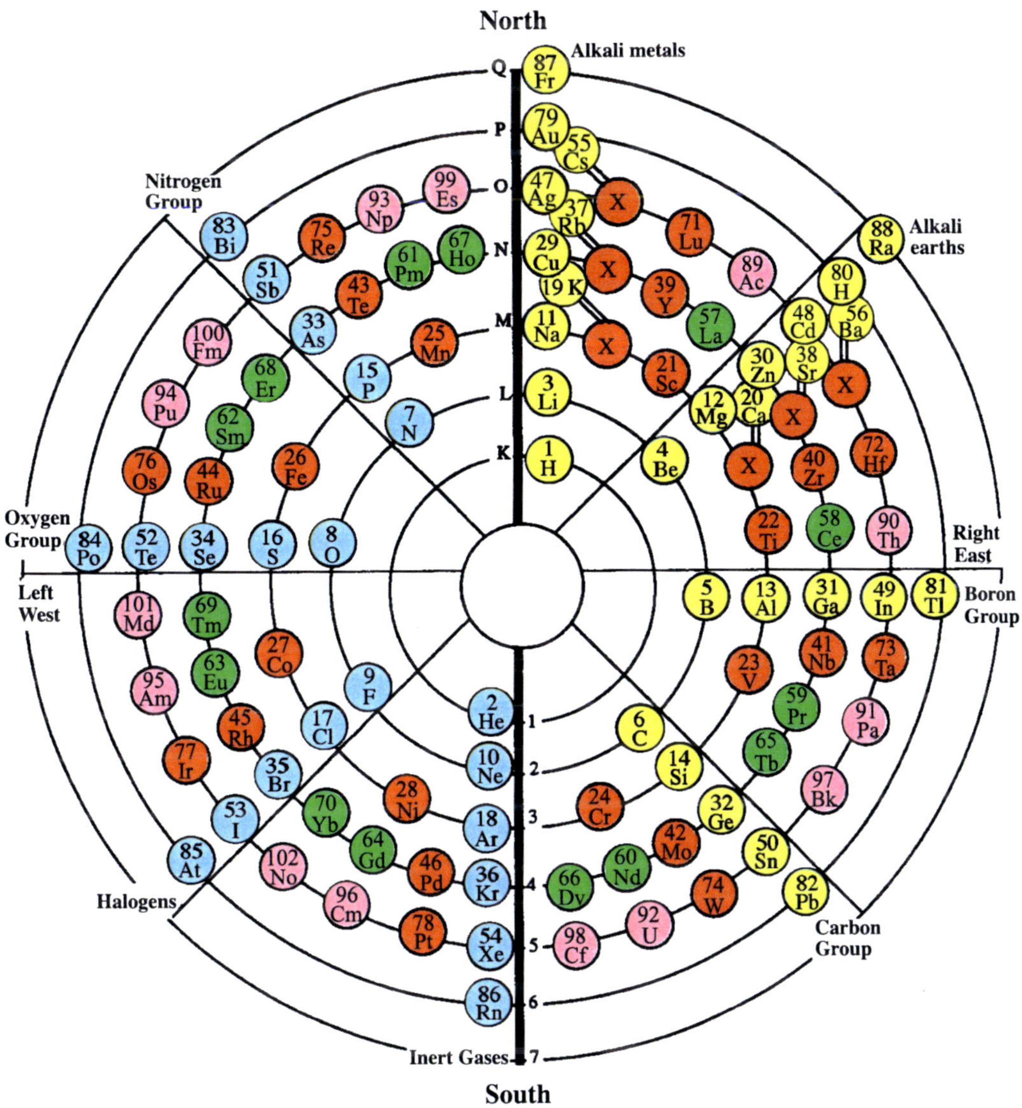

THE PAWLOWSKI MODEL OF THE ATOM

Over the past years, billions of dollars have been spent in trying to discover how the atom is structured. Yet, the answers continue to elude even the most brilliant scientific minds. Can such a problem ever be solved? I propose there is a solution, albeit a simpler, less costly one. I propose that by correcting the inconsistencies found in today's periodic table of elements, the structure of the atom is revealed. This thesis will surely be challenged, for how could something so simple have such far-reaching effects? After years of accumulating evidence, there is now ample proof to support the premise that the atom can be visualized and the periodic table corrected!

When scientists present a new theory, they give evidence of why they have come to such a conclusion. The Pawlowski Model of the Atom will now be presented in the same way. This next section will give a cursory overview of this new information in order to establish the premise upon which this book rests. Because of space limitations, it is only an abbreviated presentation. (More in-depth information can be found in Helen Pawlowski's book, *The Visualization of the Atom*.)

Readers, who perhaps feel this model of the atom is just too simple to be true, are cautioned to not be too hasty in their judgment. Many scientific theories have received the acceptance of the world when they were nothing more than theory, yet the world still clings to them as if they were fact. All that is asked is open consideration. This material does not seek to supplant existing scientific evidence, but rather to bring a different perspective to what is already known. It is a paradigm. Where there has been a lack of understanding, we feel the truths in this model will help fill the void.

Historically, new scientific theories that challenge current thought have had an extremely hard time gaining acceptance. Marilyn Vos Savant speaks of this problem: "Science is rife with error...so many people are so thoroughly schooled in the common misconceptions (that) only the most brilliantly skeptical of them will ever discover a mistake. And even then, it will likely be denied for generations to come." (*Parade Magazine*, column " Ask Marilyn," Feb. 6, 1994).

So once a theory is established within scientific thought, who dares challenge it? Hoping this doesn't sound too brash or foolhardy, meekly I say, "I do." I believe the Pawlowski Model of the Atom is correct and has the ability to fill in the missing pieces, but unless the barriers are challenged, who will ever know?

I am not here to attack the scientists, for without question, the world is a better place because of their remarkable discoveries. I am here to present what I think are the fundamental laws and principles at work in the Pawlowski Atom. But I must frankly tell you from the onset that I am not a scientific person. I will perhaps use imprecise terminology when describing scientific thought or express ideas in an unconventional way, but keep in mind I am speaking from a layman's perspective. The way I write is the way I visually see the interacting of energy within the Pawlowski Atom. Because it is the way I see it, I take full responsibility for my interpretations, expressions, and thoughts. My purpose is to get you thinking in a different way and to expose you to some core truths that science has not yet fully understood. Please bear with me in my efforts to show how they apply to life.

What are Elements, Atoms, and Electrons?

Some readers might not know what is being discussed when speaking of elements, atoms, and electrons; therefore, the following will establish a more common understanding. An *element* is one of a class of substances, which cannot be separated into substances of other kinds, or have resisted analysis by known chemical means. There are 92 of these natural forming elements. An *atom* is the smallest unit of an element. It is composed of protons, neutrons, and electrons, whose number and arrangement determine the element. An *electron* is an extremely small particle within the atoms containing a unit of negative electricity. Their positions have been uncertain.

Perhaps a simple analogy would better explain this. What is an atom? Suppose we cross an onion with a peach. This very unusual fruit/vegetable would, like the peach, have a pit or stone in the center, but like the onion, would have rings or layers surrounding its outer part. The round pit in the center would be called the nucleus, and its job would be to hold the surrounding seven circular rings to the center. Inside this pit (or nucleus) would be two types of particles: protons, which have a positive charge, and neutrons, which are neutral and have no charge. On the onion rings or shells surrounding this nucleus are more particles. These particles, called electrons, have a negative charge, and they balance out the positive energy of the nucleus.

The placement and quantity of these electrons are of prime importance, for how they build and come together determines what kind of element they become. Will they be a positive element or a negative element? Will they be a copper atom, a hydrogen atom, a gold atom, or a zinc atom, etc? As we see from the following quote, this is a most important aspect of the atom:

> The arrangement of particles that make up the nucleus of the atom is not definitely known, nor is it particularly important in general chemistry. *The arrangement of the electrons, however, is one of the most striking characteristics of the atoms of each element.* (*New Wonder World Cultural Library*, Vol. 1, p. 321, emphasis added)

Each element is made up of atoms with a distinct number of electrons and protons. They may differ in the number of neutrons. Hydrogen atoms have one electron and one proton, oxygen has eight electrons and eight protons, and calcium has 20 electrons and 20 protons, etc. By combining two or more elements, we get a compound. But remember, everything that we know of in the universe, is made of elements—that are made up of atoms—that are made up of *electrons* and protons. Atoms are the building blocks of creation, but it is the position and arrangement of the electrons that determines if an atom is either positive or negative. This is why the Pawlowski Atom is so important—it gives a visual picture of this structure. It also shows how all 92 distinguishing electrons and natural elements work together as one.

What is a Periodic Table of Elements?

It is a systematic grouping of the elements according to their similar properties. Since periodic means "a tendency to reoccur at repeated intervals," the table is supposed to reflect a repeated pattern as each element builds upon the one before it. Its purpose is to help students of chemistry and physics better understand the relationship each element has with the others. The first scientist to arrange the elements in an orderly manner was a Russian by the name of Dmitri Mendeleyev. His periodic chart is well over 100 years old and was the first to be accepted by the scientific community. Today's table, with its familiar individual squares in an elongated configuration, is an updated version of his beginning table (see page 16). But there are problems with this table that have never been resolved, such as: Why do the actinide and lanthanide series (transitionals) not fit within the main body of elements? And why are positive representative elements not seated all together? And is #1, hydrogen positioned on the right or left side of the table? And why are #29–#30, #47–#48, and #79–#80 seated with transitionals, rather than positive representative elements? Even though the table has served science well, these questions continue to confuse many.

Would science be better served if these issues were resolved? I believe so. This is what the Pawlowski Atom is all about. It has a solution to these problems. But before getting into the details of how this is done, we need to first find out how it all began.

How Did the Pawlowski Model of the Atom Come to Be?

In 1947, Helen Williams (later Pawlowski), entered a class in beginning chemistry in the small town of Spanish Fork, Utah. She was excited, although fully aware that her mathematical background was lacking. Her main interest was in the arts, so technically she was in foreign territory. But since her oil paints were made from natural elements such as cadmium (yellow), zinc (white), and cobalt (blue), she desired a basic knowledge of the earth's elements.

During the first week, a seed of unrest was planted in her mind when the teacher introduced the periodic table of elements and explained the relationship of the various elements to one another. It was the explanation of the transitionals that bothered her. She was disturbed to see the awkward positioning of these elements. Some were placed below the main body of the table, while others did not fit into the proper sequencing. These inconsistencies kindled within her an intense interest in periodic tables that was to continue throughout her life. She reviewed various chemistry books, always wanting to know more about periodic tables. She even dabbled with her own ideas of how the table should be formed.

Even though Mrs. Pawlowski studied intently from many books, it was the second hand science books of the 50s and 60s that caught her attention. These books were somewhat "outdated," and did not contain all the current theories of the day. However, since they were more elementary in their presentation of periodic tables, it was easier for her to see where problems could be found. She, of course, did not challenge the bulk of information in the field of chemistry, for her interests were always fairly restricted to the periodic table itself. But toward the end of her long years of study, another area of concern began to surface. She could see problems in the way science viewed the atom. *It was their probability/chance placement of electrons that concerned her*.

Mrs. Pawlowski wondered how there could be randomness in positioning the electrons since a highly respected chart, called Bohr's Electron Distribution Chart, has them building in a definite repeated order. It reveals how many electrons are on each shell as well as the number of electrons in each element. It also reveals the buildup of energy within the atom as well as gives a numerical description of the placement of electrons. Because it describes the atoms of each element with such great accuracy, it caused her to ask, "Would it be possible to create a model that reflected these same measurements?" Mrs. Pawlowski felt that with such precision, the structure of the atom could be visualized.

What originally had been concern over the inconsistencies found in the periodic table, she now developed an equal concern over the lack of a visual model of the atom and the need for a working relationship between Bohr's chart and the Standard Periodic Table of Elements. In 1983, she attempted to put her thoughts and ideas down in writing, but was frustrated greatly fearing they would not even be considered unless she had irrefutable proof. Strong spiritual impressions came that if she were to correct the inconsistencies of the periodic table, she would have her long sought for evidence. These impressions were so compelling that she spent all her waking hours for the next five months trying to unravel the problem. With only yarn and gold stickers as her working tools, she began to arrange the shells and electrons in many different forms and patterns. *The real breakthrough in unblocking the secrets of the atom came when she realized a polarity barrier was causing the electrons to be organized into two separate energy fields!* This new direction came into her mind more like dictation than original thought. With this knowledge, the rationale behind some unexplained movements in Bohr's chart began to make sense. When she finished arranging the periodic table in a circular configuration, to her amazement and pleasant surprise, it appeared to be the very model of the atom she had been seeking. She knew this new positioning of the electrons/elements was true to elementary chemistry because of her knowledge of the elements.

The Pawlowski Model Brings a New Perspective to Periodic Tables

A most unusual aspect of the Pawlowski Model is that it is also a periodic table of elements! This gives it credibility as the pattern of creation since combining the two into a table/model has never before been done. What is the difference between a table and a model? The table is mainly the study of chemistry (the elements) as outlined by the Mendeleyev Periodic Table while the model is mainly the study of physics (the electrons) as outlined by Bohr's Electron Distribution Chart. Niels Bohr, the originator of this chart, was a physicist from Denmark. His work has benefited science greatly because it brought the math component into the study of the atom (quantum). His chart laid the foundation for physics by numerically describing the atoms of each element. But there was a problem with his chart. Even though it showed how many electrons were on each shell as well as the number of electrons in each element, there were five areas of shifting electrons that he and his colleagues could not explain. This lack of understanding brought a probability/chance factor into the study of physics.

This is how Helen Pawlowski came into the picture. She figured out the rationale behind these irregular movements and the reason for their presence. She found, when following the buildup of electrons in Bohr's chart with exactness, that the periodic table moved from a static fixed entity to a model that has movement and life! This precise positioning of electrons in her Pawlowski model gives a dynamic, visual picture of how the atoms and elements achieve balance and equilibrium in spite of such an uneven ratio between the positive and negative. It also shows how the electrons work together as one. It is a simple, yet powerful concept. When the electrons are placed in exacting positions, the long recognized inconsistencies in the Standard Periodic Table are corrected and the structure of matter is revealed!

The Great Einstein, Bohr Debate: Exactness – Versus Probability/Chance

Albert Einstein, who was a friend of Bohr's, also disagreed with some of his conclusions. Einstein felt that in quantum theory something was missing when trying to describe minute matter. This issue caused these two noted scientists of the twentieth century to long debate the reality of the unseen world. Einstein, when speaking of the order within the elements, said, "*God does not shake dice*." Bohr and his associates held the position that there was a probability factor in the "world of the small" that made the movements and structure of matter uncertain. Bohr's theory left so many unanswered questions that those who came after him adopted the stance that there was randomness in structuring the unseen particles of the subatomic world.

Numerous writers have commented on this debate. Dietrick E. Thomsen was one of these. He stated in a *Science News* publication, "Einstein could not put up with any of this. He insisted that objects must have physical attributes that are always actual and real...He said, '*Quantum mechanics has all these uncertainties because it is an incomplete theory*. It does not tell enough because it does not know enough.'" ("As God's Dice Falls," January 1986, Vol. 129, No. 2, pg. 26, emphasis added)

John Glienman states in a June 1983 *Science Digest* article, that Einstein in a letter to a friend wrote the following (emphasis added):

> Quantum mechanics is most awe inspiring. *But an inner voice tells me this is not the real thing after all*. The theory gives much, but it scarcely brings us nearer to the secrets of the Old Man. ("Einstein Against the Odds: The Great Quantum Debate")

Abraham Pais emphasized this further when summarizing Einstein's thinking:

> "Quantum mechanics represents a major advance, *yet is only a limiting case of a theory which remains to be discovered.*" He believed that quantum mechanics would not be the starting point in the search for this missing theory. Instead, and this was Einstein's main point—*one should start all over again*, as it were and endeavor to obtain the quantum theory as a by-product of a general relativistic theory or a generalization thereof...In 1950, he wrote to Bohr: "I am convinced of (objective reality) although up until now, success is against it." (*Subtle is the Lord—The Science and the Life of Albert Einstein,* emphasis added)

Is success now at hand? Could the Pawlowski Atom accomplish the new start Einstein was speaking about? Is it the missing part of the quantum theory? At first glance the Pawlowski model might seem too simple to accomplish such an awesome task, but since it is correct and not just a good representation, I am confident that it is the missing part of quantum. This will be a major paradigm shift for most people. To discover the unique features of this pattern of creation, a person must want to look at things in a different way. Quantum will be more fully illuminated once the Pawlowski Periodic Table/Model is understood.

Modern science is built upon the measured values of quantum mechanics. The Pawlowski model provides a visual picture as to the origin of these values. It reveals the beginning and ending state of an atom once changing electrical values has impacted it. In a most remarkable way, the Pawlowski Atom connects the workings at the subatomic level to the workings of outer space. And why shouldn't it, for the structure of matter is universal. By showing how the discrete energy patterns at the minute level parallel the discrete energy patterns at the interstellar level, the pattern of creation is revealed.

Mrs. Pawlowski aligned herself to Einstein's thinking. Just as she had previously felt there were irregularities within the periodic table that could not be justified, she now also felt that because of the exact placement and distribution of electrons in Bohr's chart, the structure and dynamics of the atom could be visualized

Bohr was a Believer in Complementarity or the Opposite Nature of Things

As a side note, Bohr, the father of quantum and the defender of probability/chance, was a believer in complementarity, or the opposite nature of things (the positive and the negative). He was always impressed with the ancient Chinese notion of the polar opposites, yin and yang:

> He saw the day when complementarity would be taught in the schools and become part of general education: and better than any religion, he added, "*...a sense of complementarity would afford people the guidance they needed.*" (*Niels Bohr–Centenary Volume French and Kennedy*, pg. 323, emphasis added)

We also learn from the *Science News* article, as previously mentioned:

> ...That Bohr tried to apply complementarity to many things outside physics, even to ethics and morality. He used it for example, in discussing the paradox that human beings feel interiorly they possess free will, and yet *many things in the world go according to the laws that are deterministic and leave no room for free choice.* (January 1986, pgs 28-29, emphasis added)

How amazing! The Pawlowski Atom has the same complementary opposite structure that Bohr felt existed, but it also has the proof that God does not shake dice. It is a paradox—Einstein and Bohr were both right!

Similarity and Differences Between the Standard and Pawlowski Tables

The Standard Periodic Table and the Pawlowski Periodic Table are similar, yet different. Both contain the same basic 92 natural elements and 10 man-made elements; eight groups or families; seven circles or shells, and representative and transitional elements. Changes occur once a polarity line, electron flip, and "sliders" are added. The elements then come together in a circular configuration with all eight families in their proper positions along eight rays that are found in their proper energy fields. These three characteristics also cause the table to move from the study of chemistry to the study of physics. The table then becomes a model of all the atoms or a prototype of what is happening inside the atom. It then shows the dynamic buildup of the electrons/atoms and an organization of the positively charged and negatively charged elements as related to the polarity barrier. It also explains why the irregular movements in the upper, right quadrant are so important and how elements #29–#30, #47–#48, and #79–#80 fit in. It even solves the dilemma of where to put the transitionals.

To give you an idea on how statistically improbable it would be to produce the same exacting order as found in the Pawlowski Atom, picture in your mind's eye, throwing the 92 distinguishing electrons that make up the natural elements up in the air. Now see where they land. Could uncertainty and probability/chance have organized the electrons/elements so they would come together and do all of the above? I believe not. The Pawlowski Atom is able to show how matter is structured because it reveals where the electrons are positioned. *They are not, as previously thought, rotating around the nucleus in a circular manner*. They are instead in two different energy fields, spinning clockwise or counter clockwise in a well defined space (see page 47).

> We now know that *electrons do not orbit the nucleus* in a two-dimensional plane as planets about the sun. Orbit is a term left over from what has become an outdated view of the atom. (*National Geographic*, "Worlds Within the Atom," May 1985, emphasis added)

As you study Helen Pawlowski's table and model on the following pages, it is important to know that the circles on her *table* (page 16) represent the individual elements as found in the study of chemistry,* while the circles on her *model* (page 17) represent the distinguishing electrons making up the atoms of the individual elements as found in the study of physics.** By being able to visually connect these two fields of learning into one table/model, it creates the pattern of creation. Once its unique dynamics are understood, other fields of learning will be enhanced and refined, and, in fact, brought to a greater level of truth.

* This work will deal only with the 92 natural elements as found in Bohr's Chart.
**The model is presented as two dimensional. It is, in reality, a three dimensional sphere.

Standard Periodic Table of Elements

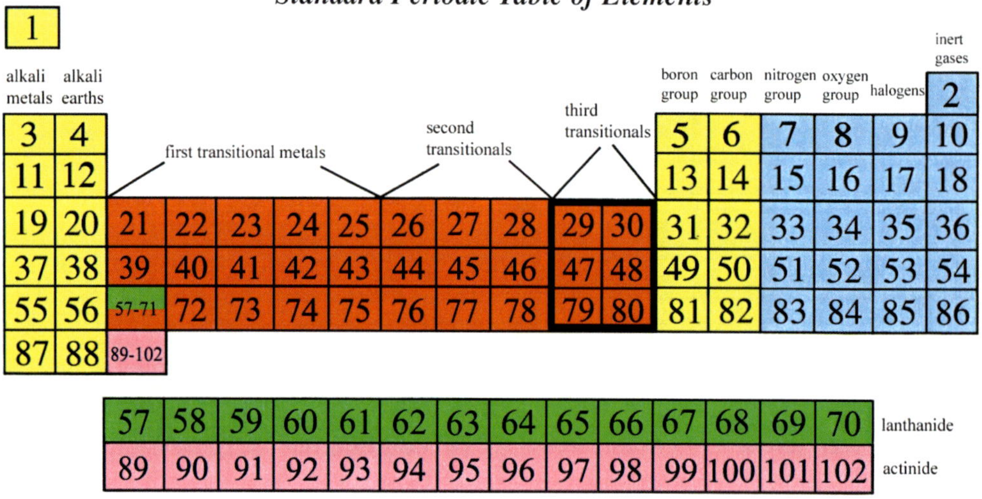

Pawlowski Periodic Table of Elements

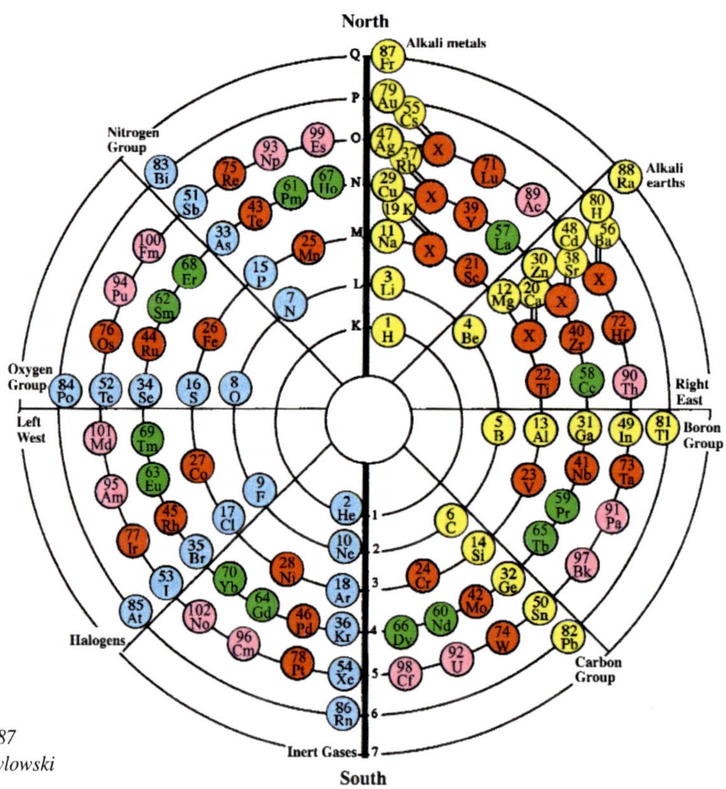

Copyright 1987
By Helen Pawlowski

The dynamics within minute matter is clarified and the opposition between positive and negative energy fields is revealed once the electrons are put in exacting positions in a circular configuration. By adding a polarity line, an electron path/reverse flip, and "slider electrons," it creates the Pawlowski Periodic Table of Elements as well as a model of all the atoms.

The Distinctive Features of the Pawlowski Model of the Atom

******

Polarity Separates Electrons into Positive and Negative Energy Fields

Negative

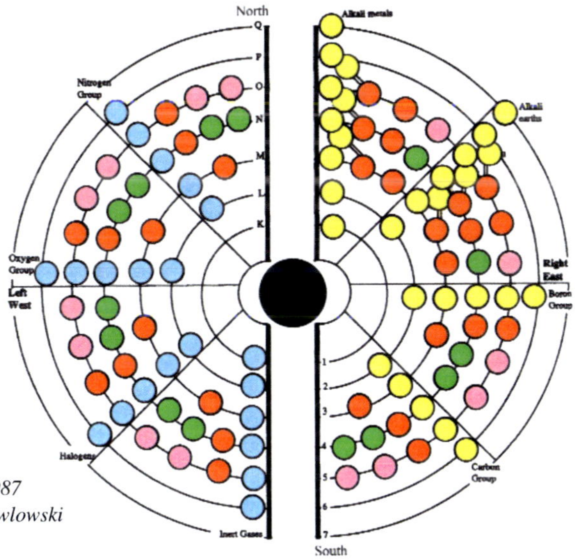

Positive

Copyright 1987
By Helen Pawlowski

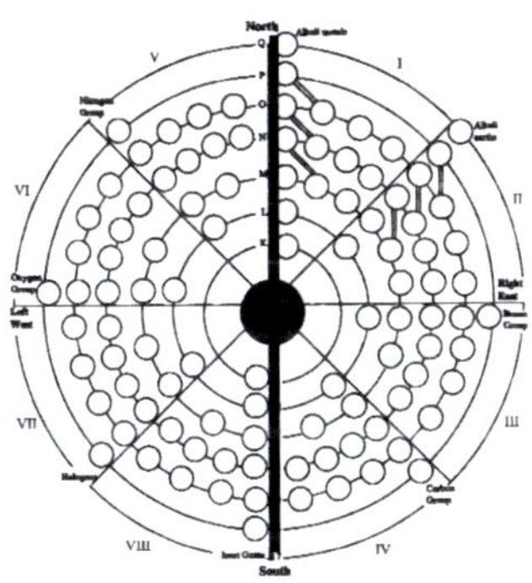

(1) North/South polarity barrier

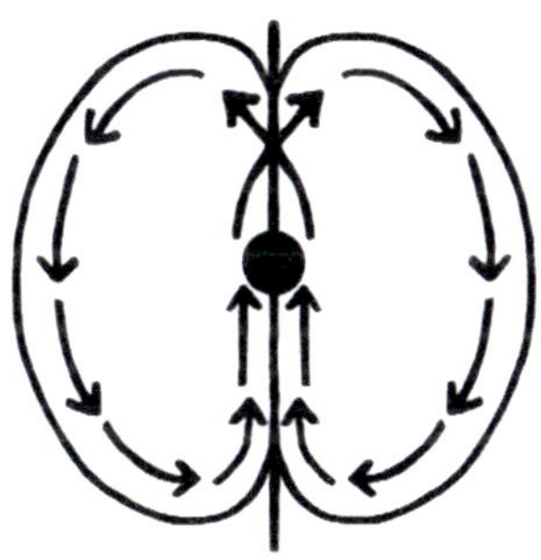

(2) figure eight electron path and reverse flip delivers the electrons

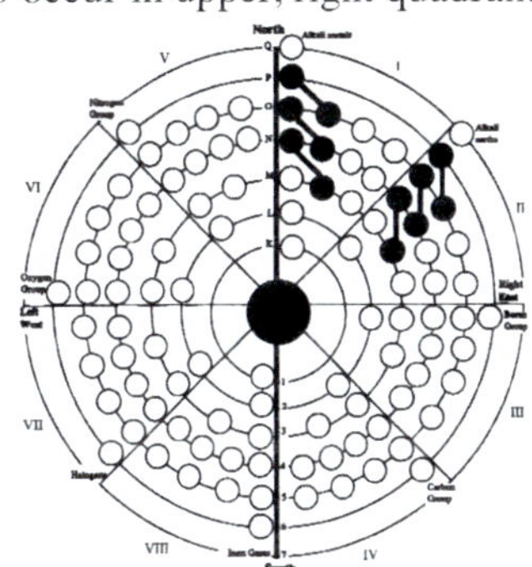

(3) "6 slider electrons" cause irregular movements to occur in upper, right quadrant

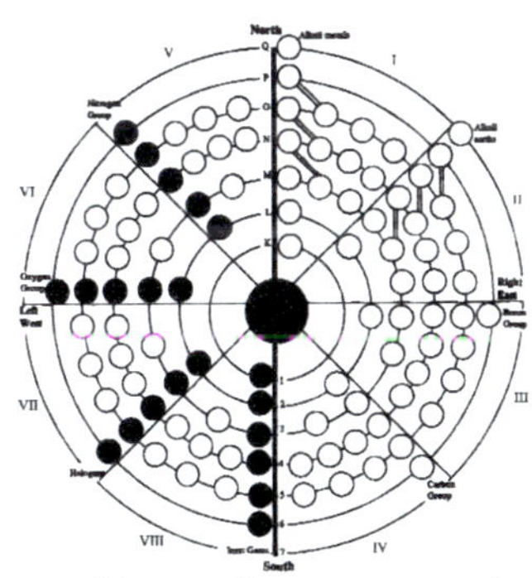

(4) 21 negative representative and inert elements on left spokes

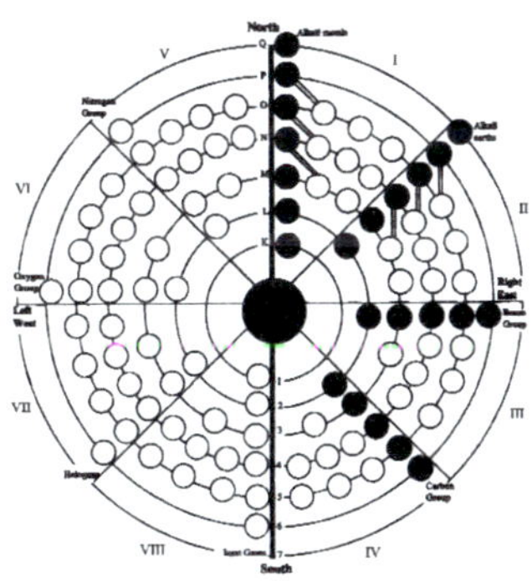

(5) 23 positive representative elements on right spokes

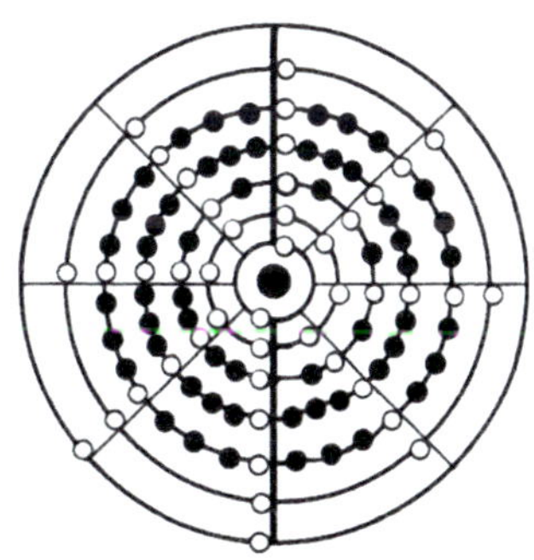

(6) 42 Transitional elements seated in inner shells – 22 to the right – 20 to the left

Exact Placement of Electrons Eliminates Probability/Chance

The Pawlowski Model of the Atom is different from what scientists have previously projected, for we see by the very positioning of the electrons that this model is governed by law, not randomness, probability, or chance. Even though at first glance it may seem too simple to merit such a judgment, those who understand its inner workings and have seen how the laws of chemistry and physics fit into its structure, are convinced such a statement is true. Yet some have asked:

> How could it be exacting, for the Heisenberg Uncertainty Principle says it is impossible to create a visual picture of the atom because when measuring an electron in one place, it disturbs an electron in another place.

This is true. However, it is important to know that this model did not come about by measurement, but by logically putting together known scientific facts. By following Bohr's Electron Distribution Chart *(see page 27)* with exactness, Mrs. Pawlowski was able to show where each electron was seated; thus creating a visual picture of what is happening within the atom.

What causes the Pawlowski model to be unique? *Its foremost feature is the north/south polarity line running through the center of the atom* **(see figure #1 on previous page)**. This dipole feature causes matter (the electrons) to divide into two energy fields: a positive field that wants to "give or share" electrons and a negative field that wants to "receive" electrons. These two halves surrounding the nucleus could be thought of as the fields around a bar magnet. This opposite arrangement gives the atom the necessary structure and components for producing both an electromagnetic and a gravitational field of energy.

Understanding the *"figure eight electron path" and the "reverse flip"* is crucial to understanding the atom **(see figure #2)**. Pretend we are building atoms and are assigned to put electrons on the shells. This buildup of electrons follows a horizontal figure eight infinity pattern. Their source (electrons) comes in at the top, which is north. The first electron is seated on the first shell, to the right, but because of the polarity barrier, the electron path cannot continue on full circle. *It can only go halfway around* (Hund's Half-filled Shell theory). It must then go back to the north and flip over the barrier so a second electron can be seated on the opposite, left side. This makes an inert gas since this shell is now complete. The same pattern is repeated on the second shell, only this time the electron path adds four electrons to the right rays before going over the barrier and adding four electrons to the left rays. This equal distribution of electrons on each succeeding shell, creates an atom that is balanced and whole.

This crossing over of the electron path verifies the existence of a polarity barrier. Since the path cannot go more than 180 degrees clockwise or 180 degrees counterclockwise before being repelled by the barrier, it must go back to the north, which adds another 180 degrees to each revolution, thus making a total 720 degrees. (This "reversal" of the electron path confirms the correctness of Feynman's Looping Belt Theory of 360 degrees to the right, then reversing 360 degrees to the left, making a total of 720 degrees.)

There are six electrons that Mrs. Pawlowski calls *"sliders"* **(see figure #3)**. *They first* appear on the fourth shell *in the upper, right quadrant. They are the signal that an adjustment must be made to the expanding atom so that balance and equilibrium can be achieved.* "Sliders" are release valves that move during the building of the atom from their positions as distinguishing electrons for six positive representative elements on the right outer shells, to six transitional positions in the inner shells. (Bohr's chart confirms that these "slider" movements and positions are not random, but exacting).

By integrating the polarity, electron path, and sliders into Bohr's chart, the electrons/elements fall into their proper positions. T*he positive and negative representative elements are seated along the eight rays projecting out from the center of the Pawlowski Atom* **(see figure #4 and #5)**. They are the "major league players" that set the perimeters of the atom. As the atom builds and expands along the rays, a balance is established when the electrons divide almost equally, with 23 to the right of the polarity barrier and 21 to the left. Those on the right east rays form positive elements that have the freedom to "act" in giving and sharing their electrons. Those on the left west rays form negative elements that must be "acted upon" by other electrons. These two energies are like a lock and a key, with one giving, the other receiving.

As the atom enlarges and expands outward along the eight rays, new space is created in the previous inner shells for the transitional elements to be seated **(see figure #6)**. They are the "minor league players." There are 22 to the right of the polarity and 20 to the left. Their seating marks the passage from a higher state to a lower state, like the modulating of keys in music. This transition first occurs on shell 4. It is there that the electron path steps back into the previous shell so the voids created by the expanding atom can be filled with transitionals. *This adjustment brings matter back into alignment and equilibrium to the developing atom*, otherwise it would wobble from lack of stability. This expansion, yet contraction of the electron path, continues throughout the remaining atoms/elements being formed. By moving from an outer shell to an inner shell, the electrons are then evenly distributed.

"Sliders" are "key players" in building the atoms because they make all the transitionals positive. By being in strong positions on the outside shell, in the upper right quadrant, they establish the positiveness or valence of the transitionals as they build on the inner shells. When "sliders" #19–#20, #37–#38, and #55–#56 finish being valence electrons for eight (or more) transitionals, they drop into the previous shell and take the position marked with an "X." By vacating their first position, #29–30, #47–48, and #79–80, which are identified as third transitionals on the Standard Periodic Table, can then take their place. These six elements have always presented a problem because they are really positive representative elements that should be positioned on the right, outside shell. Because of this unusual shifting, this problem is corrected. Many call these irregular movements unpredictable, but once you see what is happening inside the atom, the reason for such movements is clarified. (I believe these irregular movements are what science calls the Anomalous Zeeman Effect).

The buildup of electrons within the atom has a definite periodicity and rhythm that could be likened to a generator oscillating back and forth from a positive to a negative field of energy, then from a negative to a positive field of energy. Or it could be said, from a field of expansion to a field of contraction. This provides the foundation for math, with one side building energy (+), the other side taking energy (–). This vertical division of matter into separate positive and negative units assures that each atom is able to generate its own power.

The Pawlowski Atom and Bohr's Electron Distribution Chart work hand in hand in providing a visual road map to show how the structure of the atoms was not randomly put together, but was purposefully designed so that each atom could produce its own energy, as well as magnetically bring matter into line. Each in its own way tells the same thing, one with visual pictures, the other with numbers. Each reveals the unique part "sliders" or "shifting electrons" play in bringing balance and equilibrium to the whole system.

Is the Pawlowski Model of the Atom the "Theory of Everything?"

Dr. Michio Kaku, a professor of theoretical physics at City College of New York is quoted as saying in the video, *Stephen Hawkings Universe*:

> "Einstein spent the last 30 years of his life trying to create a theory of everything; a theory of black holes, of galaxies and a theory of atoms, of light, of force."

Einstein was looking for a theory that would explain the four forces governing the universe. He knew there was a strong force (the positive field), a weak force (the negative field), an electromagnetic force, and a gravitational force (the positive and negative fields working together). But what he didn't know was how they were working together as one. He felt it should be possible to unify them into one grand unifying theory. (A fifth force was later found, which I believe is the force needed for the electron path to go over the polarity barrier.)

Even though this problem continues to elude even the greatest minds, I believe that if scientists were to seriously look at the Pawlowski Atom they would find their answers, for it alone reveals how the five forces were created and how they are working together as one. It alone reveals how positive and negative matter is formed and how all things are composed of a duality of two, rather than a singularity of one. It alone reveals how electromagnetic and gravitational energy is being generated, and how "slider electrons" in the right energy field create a strong force that propels the atom onward.

The Pawlowski Atom gives scientists a new perspective in their search for truth. It makes the theory of the atom logical and capable of visualization, but best of all, it has the ability to fill in the missing pieces. Once the dynamics upon which it operates are understood, scientists will find that truth is magnified and the structure of matter revealed. Truly, the theory of everything lies within "the structure and behavior of the very smallest things," for here, the five forces of the universe were organized and the electrons positioned so they will work together as one.

The Pawlowski Atom, Like a Compass and a Clock, Reveals Our Positioning

How was our directional positioning established in the first place? Did north–south, east–west, up–down, forward–back, over–under, just evolve by themselves? I do not believe so. I believe it was embodied within matter as the atoms were being formed. Because a polarity barrier enlarged and expanded the electrons/elements into two separate energy fields as they were being created, the effects of this opposite structure has determined the nature of all things, even our universe. Since it too is made up of atoms with this opposite structure, it ultimately affects everything.

This opposition governs our space, time, and motion. According to Einstein, space and time are not separate entities, but a continuum, meaning they flow continually. I believe the Pawlowski Atom verifies this to be true, for when following the electron path (the path taken as the elements are being formed), the atom becomes like a compass that gives our position, yet like a clock, it gives our time. The twelve o'clock hour is at the top, north—the six o'clock hour is at the bottom, south. The polarity causes the three o'clock hour to be to the right, and the nine o'clock hour to be to the left. With two revolutions needed to complete the AM/PM, day/night cycle, the flow of energy is like the figure eight pattern of the Pawlowski Atom, with one side building up, the other side winding down.

The seasons also divide into four time intervals: winter, spring, summer, and fall. During one half of the year (winter to summer), the days grow longer (like the positive energy field that builds on the right side of the atom). Then at a precise moment there is a flip, and the seasons reverse into an opposite field (summer to winter), with the days getting shorter (like the negative energy field that takes away, on the left side of the atom).

Math is also based on a graph that has an x and a y-axis, x going east and west, and y going north and south. These methods, for centuries, have given us our directional bearings.

The Silhouettes of the Pawlowski Atom Show How Opposites are Formed

The silhouettes on the next page show how the elements build from the simple to the complex. They give evidence that for an atom to exist, the positive and negative must be in separate energy fields that are opposite one another. With the electrons following a right to left and left to right infinity path, this crossing over creates a positive energy field with electrons jutting out to give or share, and a negative energy field with empty spaces to receive electrons. Together, with one giving and one receiving, matter becomes balanced and whole.

One might wonder why the elements in the positive field have energy that is unrestricted and flowing out, while the elements in the negative field have empty spaces to be filled with electrons? I believe that since the positive field is the first to be created, there is no restriction on its energy because, as yet, there is no opposition from the negative side. But, once the electron path goes over the polarity into the negative field, a restriction is set up because a magnetic attraction between the positive and negative is now in place. This magnetism creates a tension between the two and like a suction, pulls the energy of the positive field back into line and leaves the negative field with empty spaces that need electrons. Because the negative field is now restricted and cannot move, its negative elements must be acted upon by positive elements.

The Silhouettes Show How the Electrons/Elements Work Together as One

As you study the silhouettes on the opposite page, you will see how the electrons/elements build one upon the other and work together as one. Hydrogen, the first element, is made up of a nucleus and one electron. By itself, it will always be hydrogen, but add another electron and helium is formed. Helium by itself will always be helium, but add another electron and lithium is formed (with the two previous electrons within its structure). On and on it goes, with each new element building upon the one that came before. Even the transitionals follow this same process. Although you cannot see the transitionals, by looking at the numbering sequence of the elements, you will notice that after "slider" elements #19–#20, #37–#38, and #55–#56, there are missing elements. These are the transitionals building on the inner shells, not along the rays. Once all the elements have been formed and seated, their electrons are working as one.

Further information suggests that the electrons are working together on an even more extended and complicated level. The illustration on page 47 confirms this statement. This visual picture gives overwhelming evidence that once an electron is seated, it remains in that position, spinning clockwise or counter-clockwise in a well defined space. It also gives evidence that each electron has a counterpart that spins in the opposite direction. With the electrons in one octet spinning up and the electrons in the opposite octet spinning down, this suggests that the alternating of positive and negative energy is what causes the electrons to spin. It also suggests that this interplay of opposite energy is what causes the positive and negative fields of the atom to also oscillate and spin. And finally, with the electrons spinning clockwise and counter-clockwise in a stationary place, where is the energy of the spins being emitted? I believe it is flowing along the same figure eight path that united the electrons, since time, space, and motion reflect this same opposite flow of energy. (Supporting evidence on pages 103–104)

These visual pictures that the Pawlowski Atom affords us, refutes the randomness that science has placed on the position of the electrons. Rather than spinning around in a "probability cloud," there is order in where they are spinning and in what direction they are spinning. If just one electron were in a different position, these silhouettes would not be possible. And if the electrons were spinning around the nucleus, the illustration on page 47 would also not be possible. It is the unique placement of the electrons, in relation to the polarity barrier, that brings stability and equilibrium to the 92 natural elements. With such stunning evidence, how can anyone doubt that the electrons were organized by exacting law, not probability/chance?

SILHOUETTES OF POSITIVE – NEGATIVE REPRESENTATIVE ELEMENTS

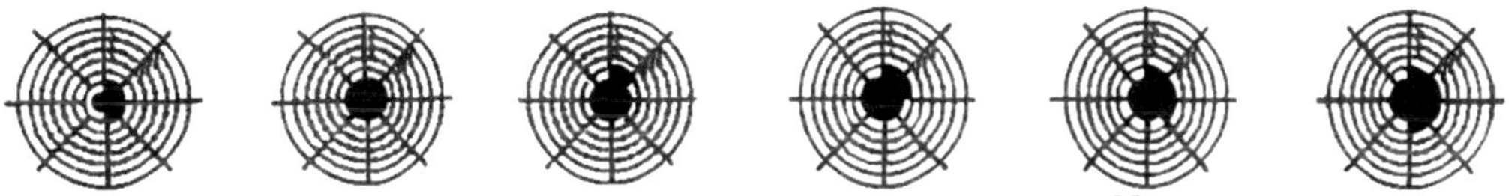

#1 Hydrogen | #2 Helium | #3 Lithium | #4 Beryllium | #5 Boron | #6 Carbon

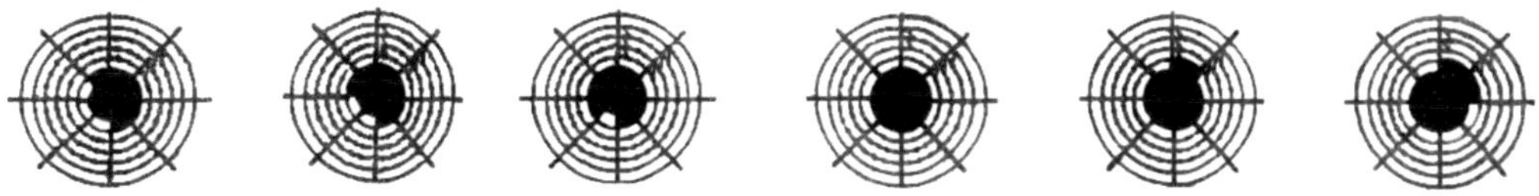

#7 Nitrogen | 8 Oxygen | #9 Fluorine | #10 Neon | #11 Sodium | #12 Magnesium

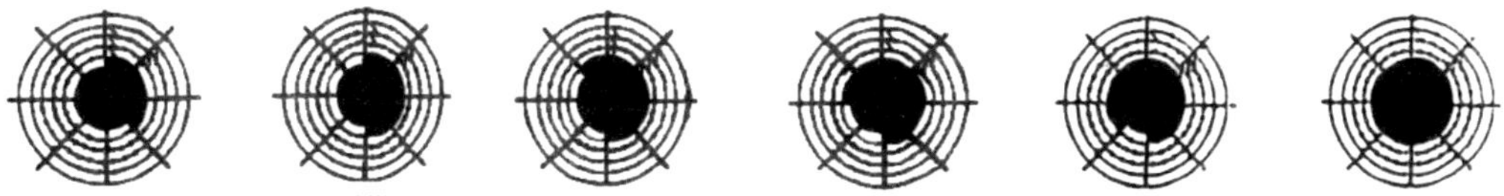

#13 Aluminum | #14 Silicon | #15 Phosphorus | #16 Sulfur | #17 Chlorine | #18 Argon

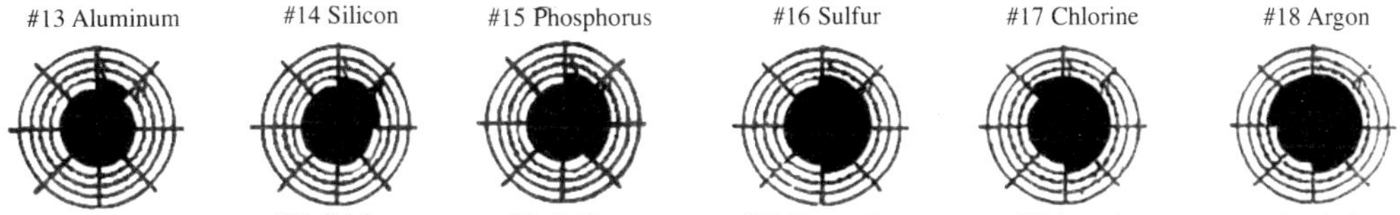

#19 Potassium | #20 Calcium | #31 Gallium | #32 Germanium | #33 Arsenic | #34 Selenium

#29 Copper | #30 Zinc

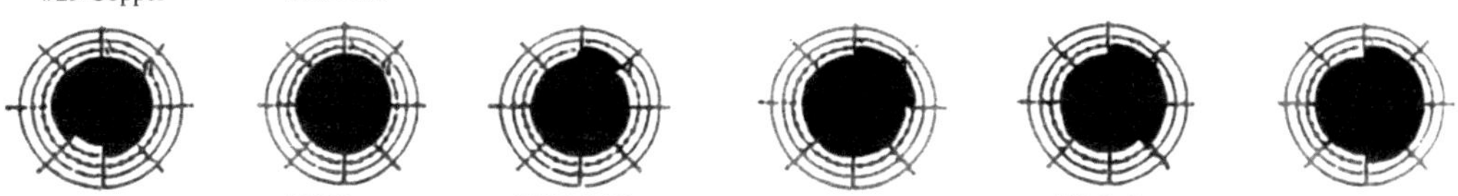

#35 Bromine | #36 Krypton | #37 Rubidium | #38 Strontium | #49 Indium | #50 Tin

#47 Silver | #48 Cadmium

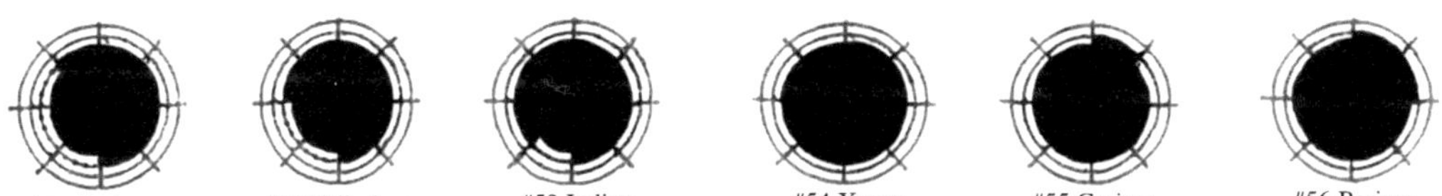

#51 Antimony | #52 Tellurium | #53 Iodine | #54 Xenon | #55 Cesium | #56 Barium

#79 Gold | #80 Mercury

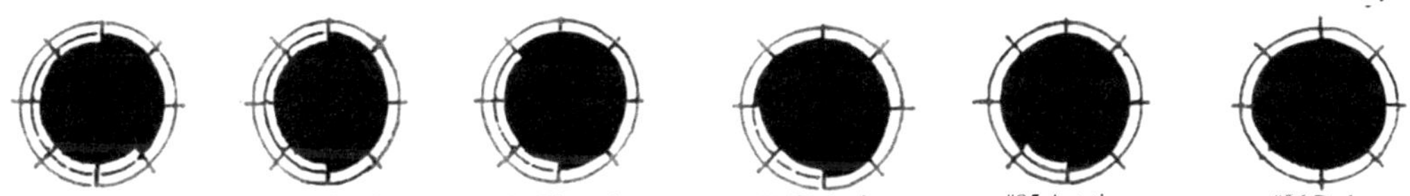

#81 Thallium | #82 Lead | #83 Bismuth | #84 Polonium | #85 Astatine | #86 Radon

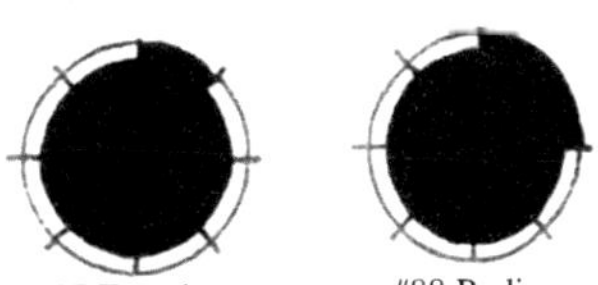

#87 Francium | #88 Radium

The transitionals are not included because they are in the interior of the atoms and do not affect the silhouettes. This is why the above elements stop at #88, rather than #92. It should also be noted that although some elements have the same silhouette, they are different because their interior rings or shells are different. These include #19–#29, #20–#30, #37–#47, #38–#48, #55–#79, and #56–#80.

THE UNITING OF BOHR'S CHART AND THE PAWLOWSKI ATOM

The uniting of Bohr's Electron Distribution Chart and the Pawlowski Model of the Atom is an extraordinary achievement. By placing the two side-by-side, it visually demonstrates how matter is governed by a law of opposites, or it could be said, "a law of balance and equilibrium." The exact placement of electrons in both the chart and the model gives overwhelming evidence that exacting law has organized matter. This statement will cause some to shrug in disbelief, since Bohr's chart has also shown in the past to have a probability factor that cannot be dismissed. How can there be two different interpretations? It is because Bohr and his colleagues overlooked some very important movements when analyzing the buildup of the electrons. Their *horizontal* picture was right, but *vertically* there were five areas of uncertainty they could not decipher. They surmised these movements played no significant part in the building of the atoms. We now know, because of the Pawlowski Atom, that these movements, in fact, play a pivotal role in determining how matter is structured and how it behaves.

Decoding Bohr's Electron Distribution Chart

On the next two pages are two versions of Bohr's Electron Distribution Chart; the one that science has used throughout the years and the one that Helen Pawlowski decoded. Although, they are the same chart, a careful examination of the two reveals an interesting pattern that Mrs. Pawlowski discovered. Once she realized a polarity barrier was bisecting the atom, she then broke the chart down vertically, rather than horizontally, into mainly 2-8-8 separate, but repeated revolutions. (A revolution means adding 8 electrons to a shell, 4 to the right and 4 to the left of the polarity.) By following these revolutions, she could see how the electrons were organized and working together as one. Horizontally, the chart still revealed the electron structure of each element, but vertically, it now revealed how the electrons, making up the 92 natural elements, were being unified in a horizontal figure eight butterfly pattern; first to the right of the polarity and then to the left. She was confident this crossing over the polarity was causing two different fields of energy to form, and that balance and equilibrium were inherent within the whole system because of this opposite buildup of electrons/elements. She also felt this alternating between a positive field that has freedom to a negative field that restricts, gave evidence that power was being generated within all the atoms/elements. Surprisingly, the chart now explained why the numbers were out of sequence in five different areas. She underlined these areas to show where uncertainty is found.

Mrs. Pawlowski found that by color coding the 2-8-8 *revolutions* in Bohr's Chart (see page 27) to correspond with the elements found in the Standard Periodic Table (see page 16)—she could more easily see where the action was taking place as the atom builds and expands. The positive and negative representative (Continued on page 28)

Bohr's Electron Distribution Chart As Used By Science

1	**hydrogen**	**1**	**47**	**silver**	**2-8-18-18-1**
2	**helium**	**2**	**48**	**cadmium**	**2-8-18-18-2**
3	**lithium**	**2-1**	**49**	**indium**	**2-8-18-18-3**
4	**beryllium**	**2-2**	**50**	**tin**	**2-8-18-18-4**
5	**boron**	**2-3**	**51**	**antimony**	**2-8-18-18-5**
6	**carbon**	**2-4**	**52**	**tellurium**	**2-8-18-18-6**
7	**nitrogen**	**2-5**	**53**	**iodine**	**2-8-18-18-7**
8	**oxygen**	**2-6**	**54**	**xenon**	**2-8-18-18-8**
9	**fluorine**	**2-7**	**55**	**cerium**	**2-8-18-18-8-1**
10	**neon**	**2-8**	**56**	**barium**	**2-8-18-18-8-2**
11	**sodium**	**2-8-1**	**57**	**lanthanum**	**2-8-18-19-8-2**
12	**magnesium**	**2-8-2**	**58**	**cerium**	**2-8-18-20-8-2**
13	**aluminum**	**2-8-3**	**59**	**praseodymium**	**2-8-18-21-8-2**
14	**silicon**	**2-8-4**	**60**	**neodymium**	**2-8-18-22-8-2**
15	**phosphorus**	**2-8-5**	**61**	**promethium**	**2-8-18-23-8-2**
16	**sulphur**	**2-8-6**	**62**	**samarium**	**2-8-18-24-8-2**
17	**chlorine**	**2-8-7**	**63**	**europium**	**2-8-18-25-8-2**
18	**argon**	**2-8-8**	**64**	**gadolinium**	**2-8-18-26-8-2**
19	**potassium**	**2-8-8-1**	**65**	**terbium**	**2-8-18-27-8-2**
20	**calcium**	**2-8-8-2**	**66**	**dysprosium**	**2-8-18-28-8-2**
21	**scandium**	**2-8- 9-2**	**67**	**holmium**	**2-8-18-29-8-2**
22	**titanium**	**2-8-10-2**	**68**	**erbium**	**2-8-18-30-8-2**
23	**vanadium**	**2-8-11-2**	**69**	**thulium**	**2-8-18-31-8-2**
24	**chromium**	**2-8-13-1**	**70**	**ytterbium**	**2-8-18-32-8-2**
25	**manganese**	**2-8-13-2**	**71**	**lutetium**	**2-8-18-32-9-2**
26	**iron**	**2-8-14-2**	**72**	**hafnium**	**2-8-18-32-10-2**
27	**cobalt**	**2-8-15-2**	**73**	**tantalum**	**2-8-18-32-11-2**
28	**nickel**	**2-8-16-2**	**74**	**tungsten**	**2-8-18-32-12-2**
29	**copper**	**2-8-18-1**	**75**	**rhenium**	**2-8-18-32-13-2**
30	**zinc**	**2-8-18-2**	**76**	**osmium**	**2-8-18-32-14-2**
31	**gallium**	**2-8-18-3**	**77**	**iridium**	**2-8-18-32-15-2**
32	**germanium**	**2-8-18-4**	**78**	**platinum**	**2-8-18-32-17-1**
33	**arsenic**	**2-8-18-5**	**79**	**gold**	**2-8-18-32-18-1**
34	**selenium**	**2-8-18-6**	**80**	**mercury**	**2-8-18-32-18-2**
35	**bromine**	**2-8-18-7**	**81**	**thallium**	**2-8-18-32-18-3**
36	**krypton**	**2-8-18-8**	**82**	**lead**	**2-8-18-32-18-4**
37	**rubidium**	**2-8-18-8-1**	**83**	**bismuth**	**2-8-18-32-18-5**
38	**strontium**	**2-8-18-8-2**	**84**	**polonium**	**2-8-18-32-18-6**
39	**yttrium**	**2-8-18-9-2**	**85**	**astatine**	**2-8-18-32-18-7**
40	**zirconium**	**2-8-18-10-2**	**86**	**radon**	**2-8-18-32-18-8**
41	**niobium**	**2-8-18-12-1**	**87**	**francium**	**2-8-18-32-18-8-1**
42	**molybdenum**	**2-8-18-13-1**	**88**	**radium**	**2-8-18-32-18-8-2**
43	**technetium**	**2-8-18-14-1**	**89**	**actinium**	**2-8-18-32-19-8-2**
44	**ruthenium**	**2-8-18-15-1**	**90**	**thorium**	**2-8-18-32-20-8-2**
45	**rhodium**	**2-8-18-16-1**	**91**	**protactinium**	**2-8-18-32-21-8-2**
46	**palladium**	**2-8-18-18**	**92**	**uranium**	**2-8-18-32-22-8-2**

Bohr's Electron Distribution Chart As Decoded By Helen Pawlowski

No.	Element	Distribution	Group
1	hydrogen	1	2
2	helium	2	
3	lithium	2-1	
4	beryllium	2-2	
5	boron	2-3	
6	carbon	2-4	8
7	nitrogen	2-5	
8	oxygen	2-6	
9	fluorine	2-7	
10	neon	2-8	
11	sodium	2-8-1	
12	magnesium	2-8-2	
13	aluminum	2-8-3	
14	silicon	2-8-4	8
15	phosphorus	2-8-5	
16	sulphur	2-8-6	
17	chlorine	2-8-7	
18	argon	2-8-8	
19	potassium	2-8-8-1	2
20	calcium	2-8-8-2	
21	scandium	2-8- 9 2	
22	titanium	2-8-10-2	
23	vanadium	2-8-11-2>	
24	chromium	2-8-13<1	8
25	manganese	2-8-13-2<	
26	iron	2-8-14-2	
27	cobalt	2-8-15-2	
28	nickel	2-8-16 2<	
29	copper	2-8-18-<<1	
30	zinc	2-8-18-2	
31	gallium	2-8-18-3	
32	germanium	2-8-18-4	
33	arsenic	2-8-18-5	8
34	selenium	2-8-18-6	
35	bromine	2-8-18-7	
36	krypton	2-8-18-8	
37	rubidium	2-8-18-8-1	2
38	strontium	2-8-18-8-2	
39	yttrium	2-8-18-9-2	
40	zirconium	2-8-18-10-2>	
41	niobium	2-8-18-12-<1	
42	molybdenum	2-8-18-13-1	8
43	technetium	2-8-18-14-1	
44	ruthenium	2-8-18-15-1	
45	rhodium	2-8-18-16-1	
46	palladium	2-8-18-18<	

No.	Element	Distribution	Group
47	silver	2-8-18-18 1	
48	cadmium	2-8-18-18 2	
49	indium	2-8-18-18 3	
50	tin	2-8-18-18 4	8
51	antimony	2-8-18-18 5	
52	tellurium	2-8-18-18 6	
53	iodine	2-8-18-18 7	
54	xenon	2-8-18-18 8	
55	cerium	2-8-18-18-8-1	2
56	barium	2-8-18-18-8-2	
57	lanthanum	2-8-18-19-8-2	
58	cerium	2-8-18-20-8-2	
59	praseodymium	2-8-18-21-8-2	
60	neodymium	2-8-18-22-8-2	
61	promethium	2-8-18-23-8-2	
62	samarium	2-8-18-24-8-2	
63	europium	2-8-18-25-8-2	14
64	gadolinium	2-8-18-26-8-2	
65	terbium	2-8-18-27-8-2	
66	dysprosium	2-8-18-28-8-2	
67	holmium	2-8-18-29-8-2	
68	erbium	2-8-18-30-8-2	
69	thulium	2-8-18-31-8-2	
70	ytterbium	2-8-18-32-8-2	
71	lutetium	2-8-18-32- 9 2	
72	hafnium	2-8-18-32- 10 -2	
73	tantalum	2-8-18-32- 11 -2	
74	tungsten	2-8-18-32- 12 -2	8
75	rhenium	2-8-18-32- 13 -2	
76	osmium	2-8-18-32- 14 -2	
77	iridium	2-8-18-32- 15 -2>	
78	platinum	2-8-18-32- 17<1	
79	gold	2-8-18-32-18-<1	
80	mercury	2-8-18-32-18- 2	
81	thallium	2-8-18-32-18- 3	
82	lead	2-8-18-32-18- 4	8
83	bismuth	2-8-18-32-18- 5	
84	polonium	2-8-18-32-18- 6	
85	astatine	2-8-18-32-18- 7	
86	radon	2-8-18-32-18- 8	
87	francium	2-8-18-32-18-8-1	2
88	radium	2-8-18-32-18-8-2	
89	actinium	2-8-18-32-19-8-2	
90	thorium	2-8-18-32-20-8-2	4
91	protactinium	2-8-18-32-21-8-2	
92	uranium	2-8-18-32-22-8-2	

elements are the yellow and blue that build along the rays. (These two colors have been reversed from a previous printing for better clarification in the slider area.) The first and second transitionals are the orange that build two shells in. The lanthanide and actinide series are the green and pink that build three shells in.

It was evident from Bohr's chart that the atom was expanding outward, like a balloon, into ever widening rays, thus creating new space for the transitionals to be seated in the previous inner shells. On the first four shells, the yellow and blue representative elements *expand* out along the rays. The path then *contracts* so the transitionals can fill in the orange inner shell. After one revolution, it shifts back to the yellow and blue representative elements along the rays. This expanding out (+) then contracting in (–), on mainly every other revolution, distributes the electrons evenly as new elements are added to the expanding atom.

She observed that whenever 2's or "slider electrons" appeared on Bohr's chart, it was the signal that new space was available in the inner shells for the transitionals to be seated. But it is interesting that after taking their positions almost half are found on the left side of the polarity barrier. You would think this would make these elements negative, but no, transitionals are always positive because the 2's or "slider electrons" on the outer shell, upper right quadrant, determine their valence. They make all the transitionals positive. This creates an uneven ratio between the positive and negative elements (71 positive, 15 negative, 6 inert gas)—but since transitionals are in the inner part establishing equilibrium, they do not upset the balance that is being created along the rays. This is how balance and equilibrium are achieved with such a preponderance of positive elements among the 92 natural elements.

Mrs. Pawlowski observed how the unexplained shifting of electrons in Bohr's chart determined where the next electrons would be seated. Also, that each electron builds on the one before it and that by adding an electron to an already existing element; an entirely new element is formed with characteristics distinct from the preceding one. If the added electron is seated on a right ray, a positive element is formed that gives or shares it's electrons, and, if seated on a left ray, a negative element is formed that receives electrons. Also, when an electron fills the last position on the bottom left ray, that shell's boundary is complete and an inert gas is formed.

Balance and Equilibrium Within the Atom

Mrs. Pawlowski could see that because of repeated patterns in Bohr's chart, balance and equilibrium are inherent within the expanding atom. As evidence that balance is important, the first two electrons are positioned on both sides of the polarity barrier, thus creating two legs to build on, rather than one. This equal distribution of matter continues throughout the building of the atom. Because of the polarity, the 8's do not go full circle, but rather split into 4's so that balance can be established between the two energy fields being formed. *By shell 4, this balance is at risk.* The atom has expanded to the point where the new space created in the previous inner shell, must be filled with electrons, or the atom will lose its stability. The 2's (with the exception of the first 2*)* signal that a correction will be made and the numbering sequence disrupted. The 2's are the "slider electrons" that shift positions in the underlined areas of Bohr's chart. If you look closely, the small arrows point to *where* and *when* the "sliders" drop into the previous shell. No one, including Bohr and his associates, could figure out why the orderly buildup of the electrons suddenly became so irregular. It is because, at this point, the transitionals start filling the enlarged inner shells so that equilibrium can be established.

Equilibrium is important to the expanding atom, otherwise it would vibrate and quiver. Equilibrium is an *inside* phenomenon and balance an *outside*. Balance comes from the electrons on the *outer* rays being of near equal strength, while equilibrium from the filling in of electrons in the *inner* shells. This same opposite structure is reflected in our own bodies. Like the eight outer rays of the atom being balanced, our two arms and two legs with their 10 fingers and 10 toes are balanced to keep us in a stable, upright position. However, if someone were to say, "I'm losing my equilibrium," we would know immediately that something *inside* his or her body was not right. Their *external* balance would still be there, but their *internal* equilibrium would need fixing. Without it, their body would remain unstable. This is why the inner spaces of the atom must be filled with transitionals as it enlarges and expands along the eight outer rays. It corrects the imbalance brought on by the atom getting larger.

How balanced are the electrons in the Pawlowski Atom? With 92 natural elements to account for, there are 23 positive representative elements on the right rays and 21 negative representative elements on the left rays. In the inner shells, there are 22 transitionals seated to the right of the polarity and 20 seated to the left. This makes a total of 86 electrons/elements. The missing six electrons/elements are the six "sliders" that shift from positive representative positions on the outer shells to transitional positions in the inner shells.

Sliders: The Cause of Probability/Chance

The numbers in Bohr's Electron Distribution chart are very predictable until coming to the areas where the elements are underlined. The numbering sequence then changes from being predictable to unpredictable. It is understandable why probability/chance entered the mindset of the scientists, for how else could the scrambling that started with scandium be explained? The numbers on this inner shell read #9, #10, #11, #13, #13, #14, #15, #16. We now know that these numbers are out of sequence because they reflect that a "slider" has dropped down into this inner shell during this revolution. But as you look on the outer shell, the numbers there also reflect that this "slider," after dropping down and being counted, went right back up to its original position. Given that this is the first of the "slider" movements, this unusual move reflects that the atom is not quite ready for transitionals to be seated (or equilibrium to be established) since it is the only time that a "slider" goes back up after dropping down. Not until the end of this revolution do "sliders" #19 and #20 drop into the inner shell, which is reflected by the numbers going from #16 to #18 at the beginning of the next revolution. To help you see where these unusual "slider" movements occur when building a model of all the atoms, a fan-like illustration will be at the side of elements #23, (#24), #28, #40, #45, #77, and #78, so that you will know where "sliders" #19–#20, #37–#38, and #55–#56 drop into transitional positions.

To my thinking, "sliders" are the missing pieces of the puzzle. I also believe that because of this shifting process, a slight irregularity is being reflected in all scientific endeavors and that this anomaly continues to be explained as a probability/chance phenomenon. These movements are so small, that it is easy to dismiss their importance. However, by knowing the purpose of "sliders," everything falls into place. I believe they are like release valves that control the pressure of the expanding atom. Remember how Rick Gore said, "In the structure and behavior of the smallest things, lies the explanation of all matter." I believe this is true, for the balance and equilibrium that is manifest throughout the universe is made possible by the smallest electrons being positioned in their proper place.

The Visualizing of the Pawlowski Atom

This work is about painting a visual picture in the mind. It is about visualizing the structure of the atom as it builds and expands. With the current pictures of the day showing the electrons moving around the nucleus in a random, probability cloud, one cannot visualize their positions. This absence of structure and boundaries does not conjure up in the mind that law is governing their movements. The Pawlowski Atom presents a new and exciting picture. It reveals that the law of opposites (embodying the law of balance and equilibrium) is organizing the electrons into two energy fields that are opposite one another—one with a positive charge and one with a negative charge.

To better visualize this buildup of electrons, one must realize that rhythm and movement are part of the process. I found this to be true when first trying to understand the Pawlowski Atom. My mind just couldn't seem to comprehend or visualize how the static, fixed elements were working together as one. There seemed to be no relationship or connection until I started putting the 2-8-8 pattern of the electron path to a musical beat. With the mind and body then working together, things began to make sense and I could visualize the buildup of electrons/elements in a new and different way. It was much like a Greek dance with all the participants dancing in two circles, interlocking arms, and systematically dropping electrons as they swayed first to the right circle (dropping four electrons) then crossing over and swaying to the left circle (dropping four electrons). I found that by visualizing and feeling the rhythm and movement of the figure eight path, I could then understand that *it was the electrons making up the atoms of each element that were working together as one, not the individual elements by themselves*. What a simple solution to a perplexing problem!

Mrs. Pawlowski's work brings order and cohesiveness to ones thinking of the atom. How appropriate that she would name her book, *The Visualization of the Atom*!

THE BUILDING OF THE PAWLOWSKI ATOM

We will now build a model of all the atoms by following Bohr's Electron Distribution Chart with exactness. The end result will be the Pawlowski Model of the Atom. They are separate parts of the same road map, so we will use both in clarifying how the atoms are formed and structured—one through numbers and one through visual pictures. This will be a departure from what has been done in the past, since many have thought that only a mathematical description could describe the atoms. You will find this new model is aesthetically appealing because of the simplicity and harmony that exists between its various parts. It resolves many issues, but most importantly, it gives striking evidence that the "structure and behavior" of the atom was brought about by exacting law, not probability/chance.

To build a model of all the atoms, we will follow the same path the electrons take as new elements are being formed. This is why it is called the "*electron path.*" I also refer to it as the infinity path since its horizontal figure eight pattern has no beginning or end. To become familiar with this path, follow the numbers in the center of each element on the next ten pages, and you will feel the rhythm of the atoms building and twisting in an upward spiral from shell to shell like the double helix spiral of DNA.

When following this path, the elements naturally fall into their proper positions. The positive and negative representative elements set the boundaries along the rays. The first and second transitionals and the lanthanide and actinide series fill the inside shells as the atom builds and expands. The 2's or "slider electrons" drop at given points from their positions on the outside shell to positions on the inside shell. The third transitionals fill the space that the "sliders" leave when dropping into the inner shell. When the path *goes from a lower shell to a higher shell*, positive and negative representative elements are formed. But when 2's or "slider electrons" appear, the path shifts and *goes from a higher shell to a lower shell* so transitional elements are formed.

As we build a model of all the atoms/elements on the subsequent pages, notice how the electrons are seated, first to the right of the polarity then to the left. In this way, each electron has a counterpart in the opposite energy field. Also notice that when an atom/element is separate and alone, its electrons function only as a part of that atom/element; but when another electron is added, the previous electrons then become part of the next atom/element. For example: Hydrogen, the first element, is made up of a nucleus and one electron. By itself, it will always be hydrogen, but add another electron and helium is formed. Helium by itself will always be helium, but add another electron and lithium is formed. By the time uranium, the last of the 92 natural elements, is formed, all the previous electrons are now part of that element.

As each new atom/element is being formed, you will notice there is a large black circle at the beginning. This is the *distinguishing electron*. It distinguishes the newly formed element from all the others. The small gray circles are the electrons from the previous elements that are now a part of the new element. The white circles are where future electrons will be seated. The black centerline represents the polarity barrier.

In setting up these visual pictures of the atoms, the 2-8-8 pattern that runs throughout Bohr's chart is emphasized by placing the 2's on the left-hand pages and the 8-8's on the right-hand pages. This makes the electron path easier to follow since the 2-8-8 pattern is in full view.

The Step By Step Building of the Atoms

Step 1. The first atom or positive element to be formed is #1, hydrogen. It's distinguishing electron is positioned to the right of the polarity, on the 1st shell, north ray, in the positive energy field. Follow the figure 8 electron path over the barrier and place the next distinguishing electron on the opposite south ray, in the left negative energy field. The first electron then becomes part of the second element. This shell's boundary is now complete, so Helium, #2, is an inert gas.

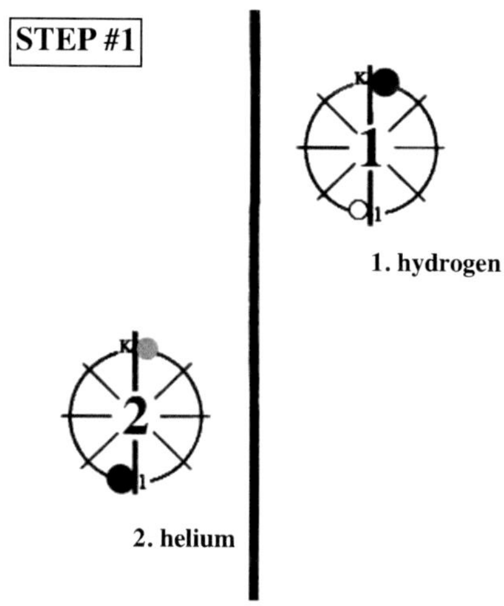

Step 2. (Top opposite page) Back to the top, north and add four electrons to the right rays, 2nd shell, and positive elements #3, #4, #5, and #6, are formed. Then back over and add four electrons to the opposite left rays and negative elements #7, #8 ,#9, and #10 are formed. Since this shell's boundary is complete, Neon, #10, is an inert gas.

Step 3. (Bottom opposite page) Add four electrons to the right rays on the 3rd shell and positive elements #11, #12, #13, and #14 are formed. Add four electrons to the opposite left rays and negative elements #15, #16, #17, and #18, are formed. This shell's boundary is complete, so argon #18, is inert gas.

(The filling of one-half of the shell before *crossing* over *the polarity* and filling the other half reflects Hund's half-filled shell theory. Also, the going forward then coming back of *the figure eight electron path* reflects Feynman's Looping belt theory.)

Eight Electrons Build on Both the 2nd and 3rd Outer Shells

STEP #2

3. lithium

4. beryllium

5. boron

6. carbon

7. nitrogen

8. oxygen

9. flourine

10. neon

STEP #3

11. sodium

12. magnesium

13. aluminum

14. silicon

15. phosphorus

16. sulphur

17. chlorine

18. argon

Two Electrons on 4th Outer Shell

Step 4. Next are #19 and #20, the first of six "sliders." They are the signal that the atom has expanded to the point where new space is available for the transitionals to be seated in the previous, inner shell. This is the first two-electron shell in which both distinguishing electrons are seated to the right of the polarity barrier. Later, these two electrons (#19 potassium and #20 calcium), slip down into the shell closer to the nucleus and become part of the "d" series. When this happens, a small fan like illustration will be next to the elements involved.

STEP 4

19. potassium
2-8-8-1

20. calcium
2-8-8-2

Step 5. (Top opposite page) Although "sliders" #19 and #20 start on the 4th shell, to the right, between the making of elements #23 and #24; "slider" #20 drops down into shell 3. But as you can see, it immediately goes back up between #24 and #25. I believe this unusual movement has to do with the equilibrium that is soon to be established within the atom. Since it is not quite ready for this movement to occur, it goes right back up. After "sliders" #19 and #20 set the valence for transitionals #21, #22, #23, #24, #25, #26, #27, and #28 in the inner shell, they then drop into the 3rd shell between the making of elements #28 and #29, thus making their former positions on the 4th shell available. Can you now see how #29–30, #47–#48, and #79–#80 fit in? They take the place that "sliders" #19–#20, #37–#38, and #55–#56 previously had. Their positions on the standard periodic table have always been uncertain because they are labeled as third transitionals, but are really positive representative elements. The "sliders" and their replacements both have the same valance (+1) and ground state number (s1/2). Both have the same azimuthal value (s=0) and quantum principle number (potassium/copper is 4, rubidium/silver is 5, and cesium/gold is 6).

Step 6. (Bottom opposite page) After #29–#30 have taken the place that "sliders" #19–#20 previously had, #31 and #32 are added to the next two rays. Then go over the polarity and add #33, #34, #35, and #36, to the opposite left rays. This shell's boundary is complete, so Krypton #36, is an inert gas.

Eight Electrons Build On 3rd Inner Shell And 4th Outer Shell

STEP 5

21. scandium
2-8-9-2

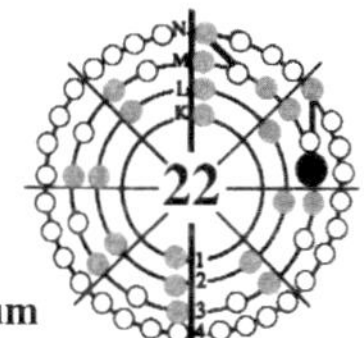

22.titanium
2-8-10-2

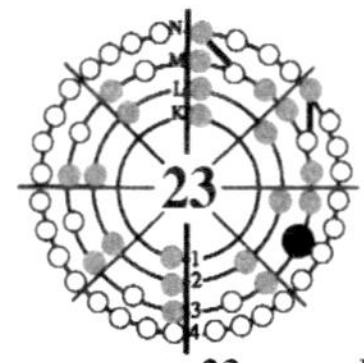

23. vanadium
2-8-11-2

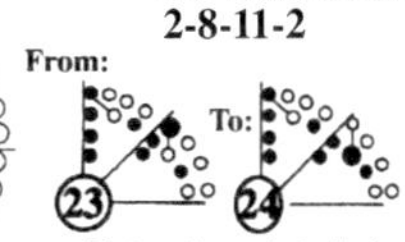

(Note enlarged circles)

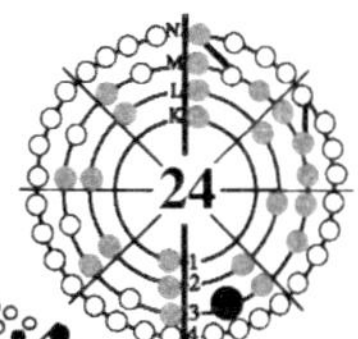

24. chromium
2-8-13-1

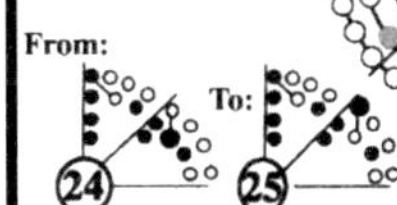

(Note enlarged circles)

25. magnesium
2-8-13-2

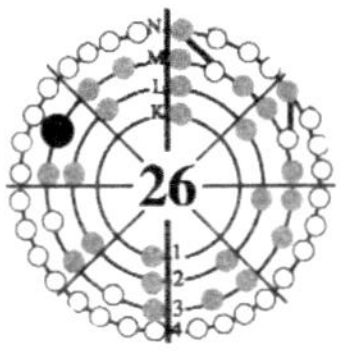

26. iron
2-8-14-2

27. cobalt
2-8-15-2

28. nickel
2-8-16-2

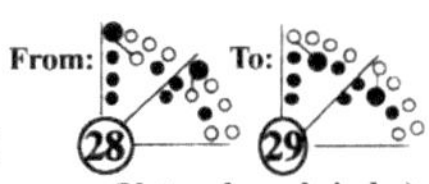

(Note enlarged circles)

STEP 6

29. copper
2-8-18-1

30. zinc
2-8-18-2

31. gallium
2-8-18-3

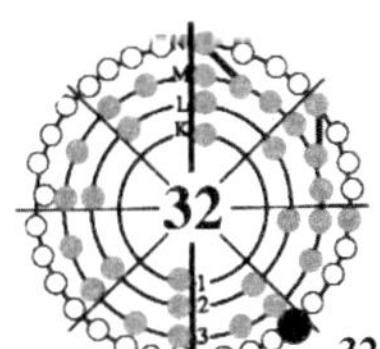

32. germanium
2-8-18-4

33. arsenic
2-8-18-5

34. selenium
2-8-18-6

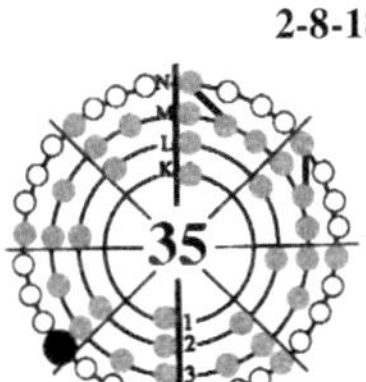

35. bromine
2-8-18-7

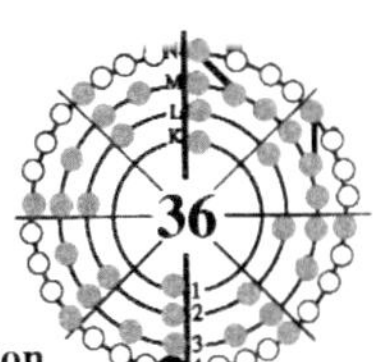

36. krypton
2-8-18-8

Two Electrons on 5th Outer Shell

Step 7. Go back to the north and add "sliders" #37 and #38, to the 5th shell.

Step 8. (Top opposite page) With the "sliders" in the upper right quadrant now in place, the electron path drops into the 4th inner shell for transitionals #39, #40, #41, #42, #43, #44, #45, and #46 to be seated. Bohr's chart then tells us that between the making of elements #40 and #41, "slider" #38 drops into the 4th shell, thus leaving elements #41 through #45 with only one valence electron. The remaining "slider," #37, drops into the previous shell between the making of #45 and #46, thus causing #46, palladium, to be more of an inert element than a positive one since there are now no electrons on the right side setting its valence. This unusual movement occurs halfway through the forming of the 92 natural elements.

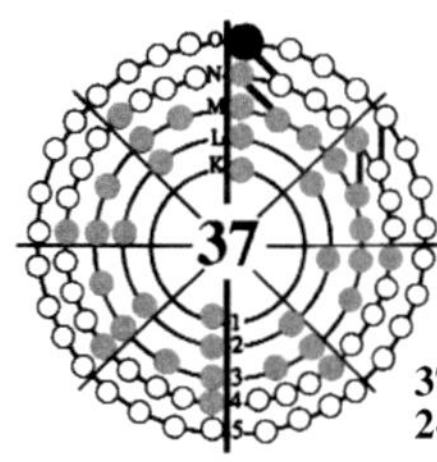

STEP 7

37. rubidium
2-8-18-8-1

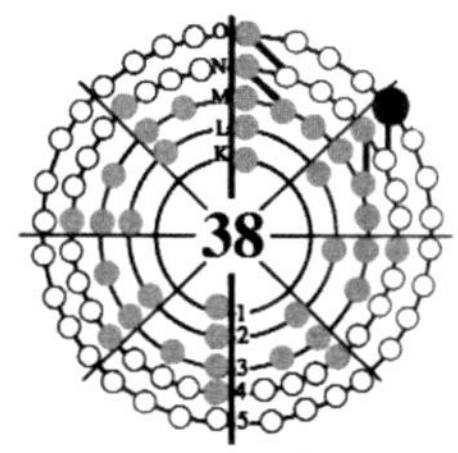

38. strontium
2-8-18-8-2

Step 9. (Bottom opposite page) Since "sliders" #37 and #38 have now dropped to the 4th shell, #47 and #48 fill their positions. Continue with #49 and #50 along the next two rays. Then over the barrier and add #51, #52, #53, and #54 to the opposite left rays. This shell's boundary is now complete, so #54 Xenon is an inert gas.

Can you now see the importance of having the electrons positioned in opposite positive and negative energy fields? It creates power, magnetism, energy, and light, or in other words, an electrical world that is governed by law not probability chance. It also creates an opposition in all things.

Eight Electrons Build on 4th Inner Shell and 5th Outer Shell

STEP 8

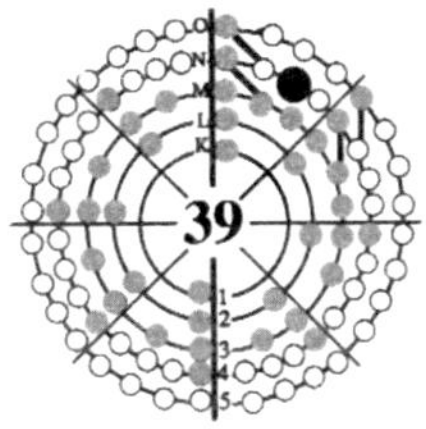

39. yttrium
2-8-18-9-2

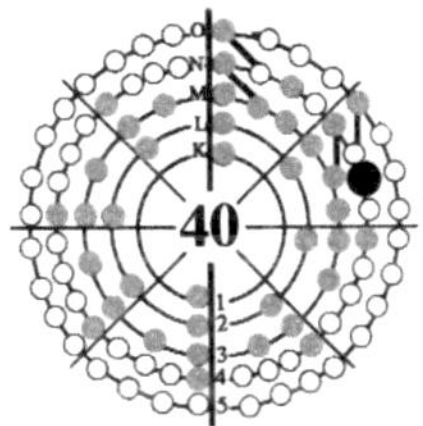

40. zirconium
2-8-18-10-2

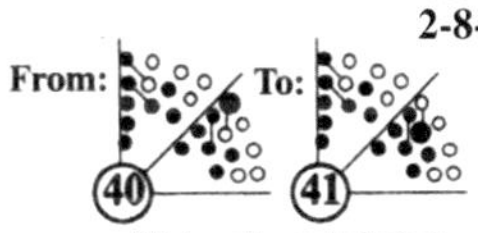

(Note enlarged circles)

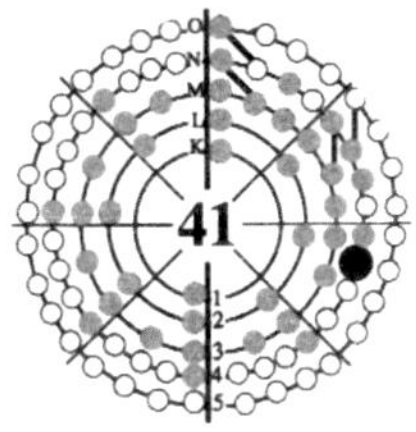

41. niobium
2-8-18-12-1

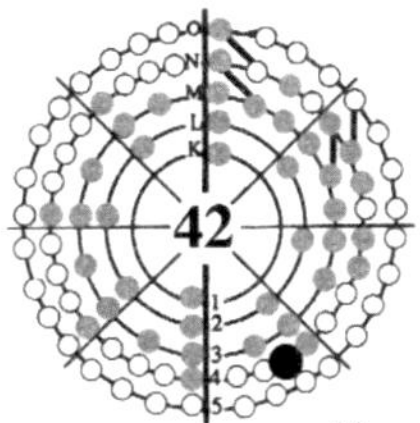

42. molybdenum
2-8-18-13-1

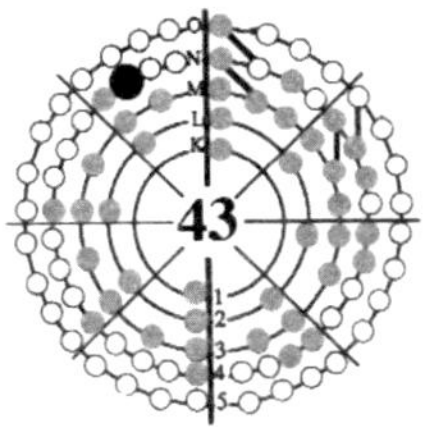

43. technetium
2-8-18-14-1

44. ruthenium
2-8-18-15-1

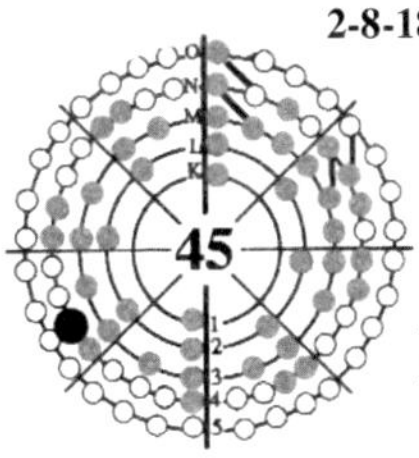

45. rhodium
2-8-18-16-1

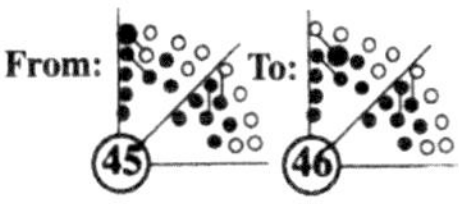

(Note enlarged circles)

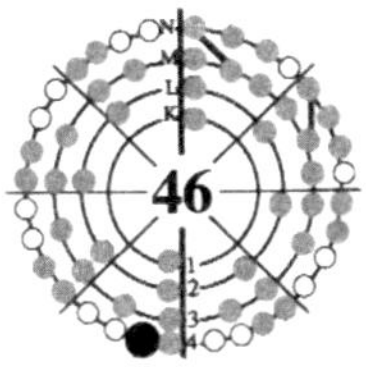

46. palladium
2-8-18-18

STEP 9

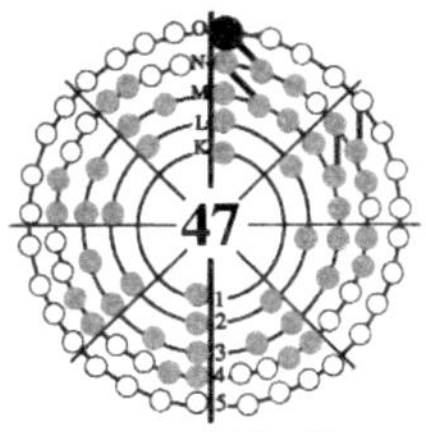

47. silver
2-8-18-18-1

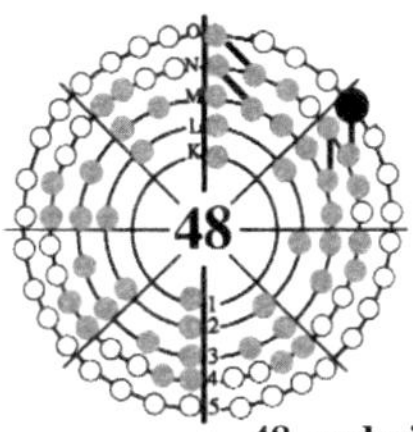

48. cadmium
2-8-18-18-2

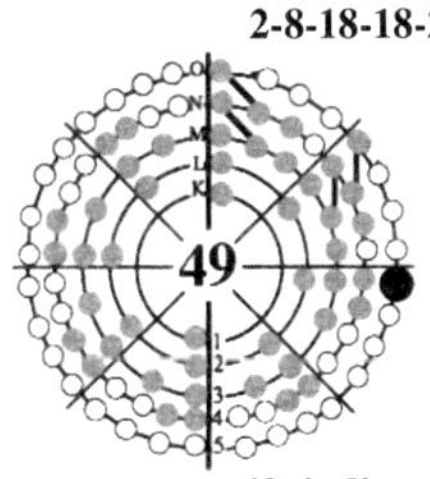

49. indium
2-8-18-18-3

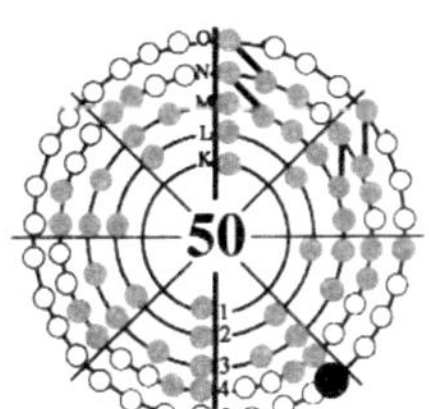

50. tin
2-8-18-18-4

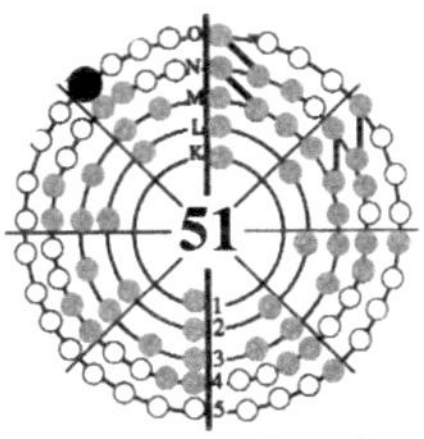

51. antimony
2-8-18-18-5

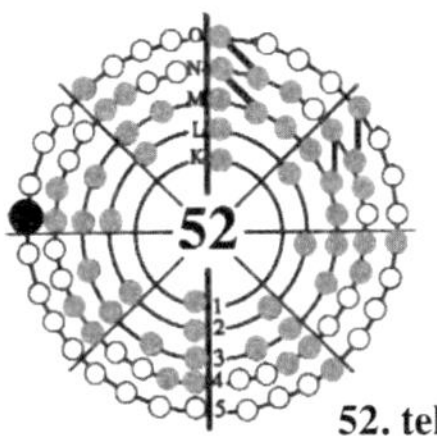

52. tellurium
2-8-18-18-6

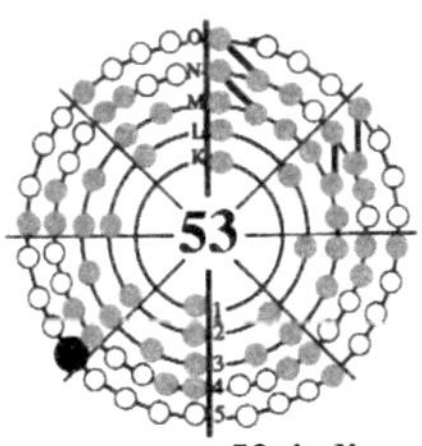

53. iodine
2-8-18-18-7

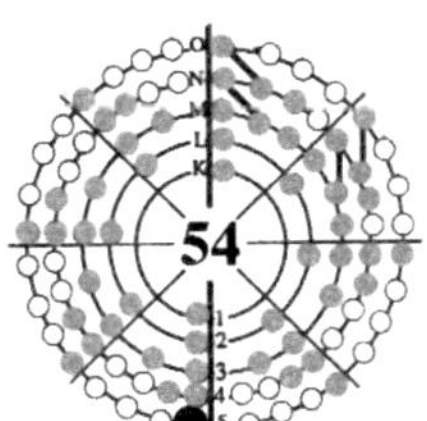

54. xenon
2-8-18-18-8

Two Electrons Build on 6th Outer Shell

Step 10. Add "sliders" #55 and #56 to the right on the 6th shell.

Step 11. (Opposite page) At this point in Bohr's Chart, the electron path drops two shells in, rather than one, so that transitionals, #57, #58, #59, #60, #61, #62, #63, and #64 can fill the expanded inner space on shell 4. After one revolution, the shell is still not filled, therefore, the electron path continues around a second time so that transitionals #65, #66, #67, #68, #69 and #70 can fill the remaining space. Notice how on the second revolution, the first two octets are filled. This is because previous "slider electrons" have dropped into this space. Therefore, #65 and #66 take the next available space in octets 3 and 4. This going around the shell a second time is why the sequence in Bohr's chart reflects a change from a periodicity of 8 to one of 14. Add the two filled octets and it makes 16, with both revolutions having a balanced 8.

STEP 10

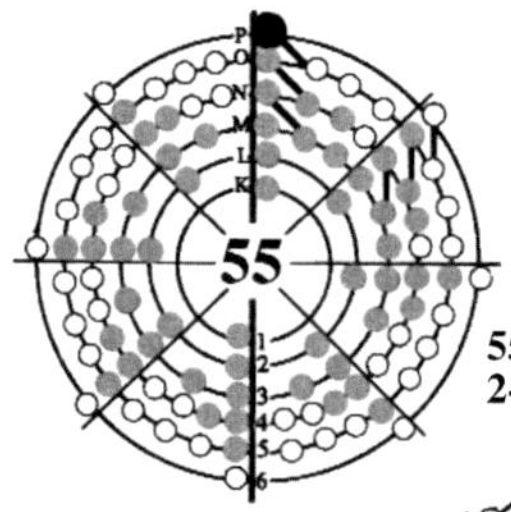

55. cesium
2-8-18-18-8-1

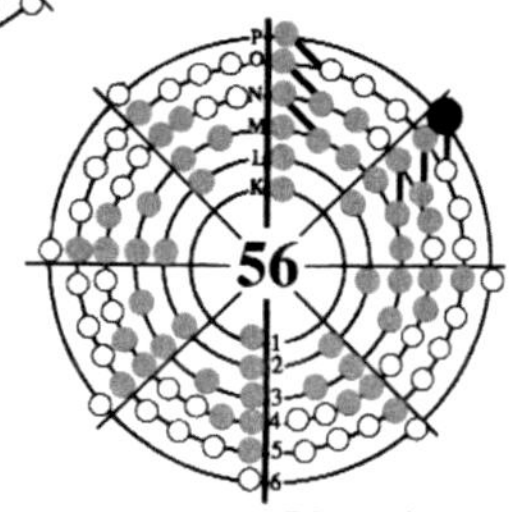

56. barium
2-8-18-18-8-2

Step 12. (See page 40) With shell 4 now complete, the electron path goes up to shell 5 so that transitionals #71, #72, #73, #74, #75, 76, #77, and #78 can fill its expanded inner space. Once these three revolutions are complete, "sliders" #55 and #56 drop down into shell 5, (Bohr's chart tells us that this occurs between the forming of elements #77 and #78, as well as #78 and #79).

Step 13. (See page 40) The two "slider" positions on the 6th shell, right rays, are now available for #79 and #80 to be seated. Continue with #81 and #82 on the next two rays. Then over the barrier and place #83, #84, #85, and #86 on the opposite left rays.

Can you now see how seating the transitional elements in the inner shells solves the problems found in the Standard Periodic Table? The actinide and lanthanide series now fit within the main body of elements and elements #29–#30, #47–#48, #79–#80 are now positive representative elements rather than third transitionals. This new structuring of the atom also helps us understand why hydrogen, the first element, is positioned on the right side of the periodic table.

Fourteen Electrons Build on 4th Inner Shell – Twice Around

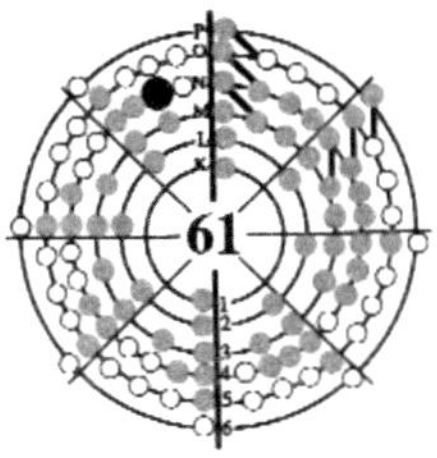

61. promethium
2-8-18-23-8-2

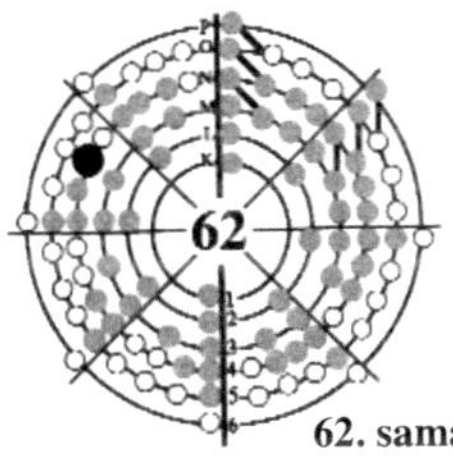

62. samarium
2-8-18-24-8-2

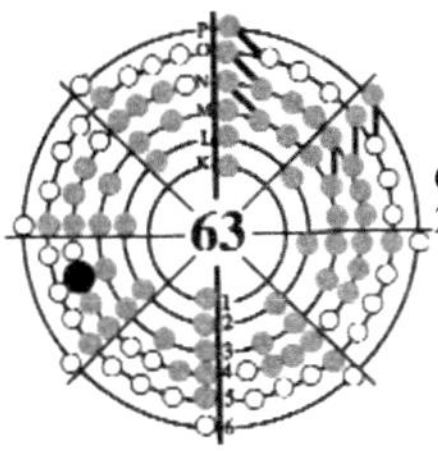

63. europium
2-8-18-25-8-2

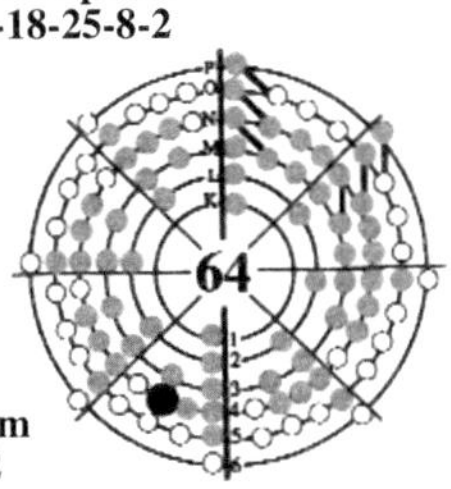

64. gadolinium
2-8-18-26-8-2

STEP 11

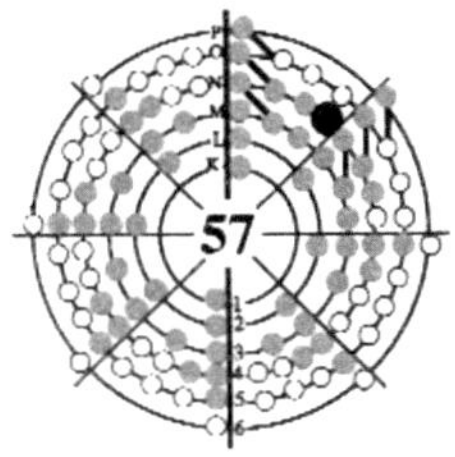

57. lanthanum
2-8-18-19-8-2

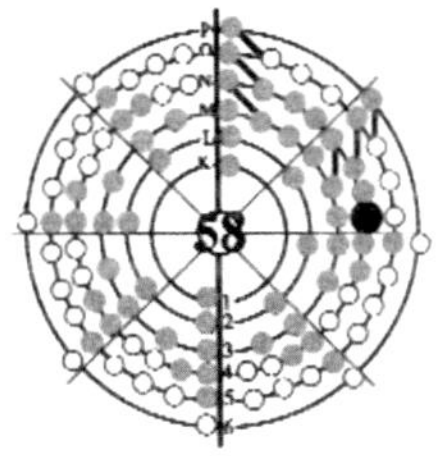

58. cerium
2-8-18-20-8-2

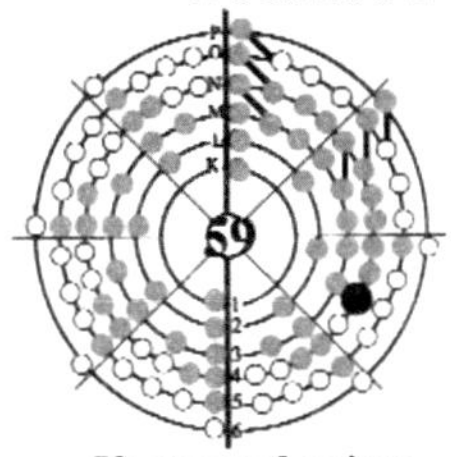

59. praseodymium
2-8-18-21-8-2

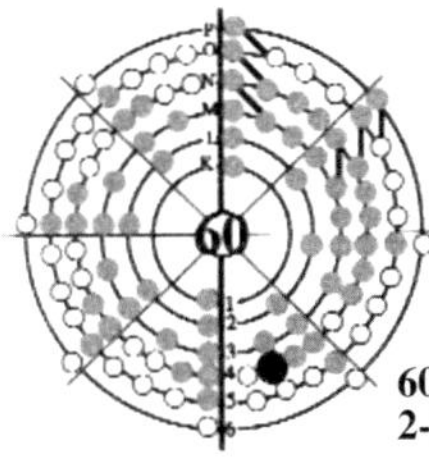

60. neodymium
2-8-18-22-8-2

67. holmium
2-8-18-29-8-2

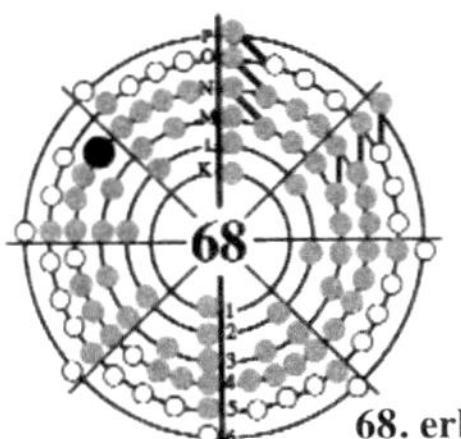

68. erbium
2-8-18-30-8-2

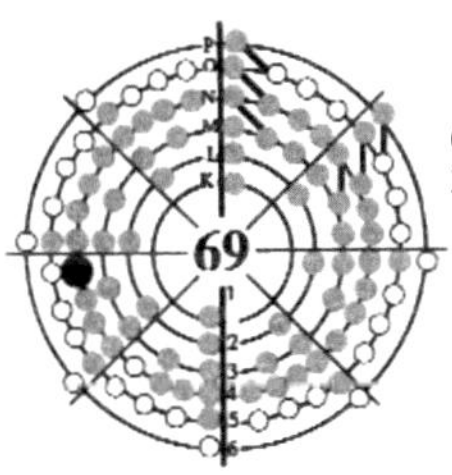

69. thulium
2-8-18-31-8-2

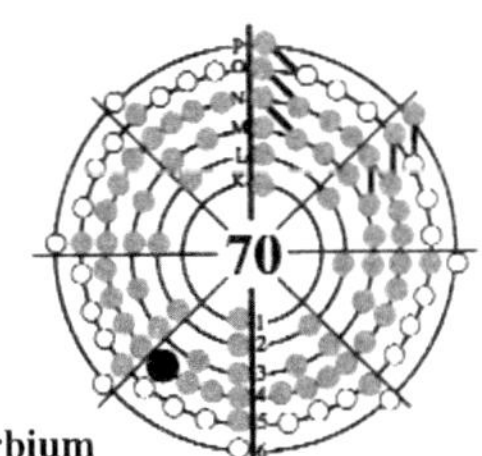

70.ytterbium
2-8-18-32-8-2

STEP 11
Continued

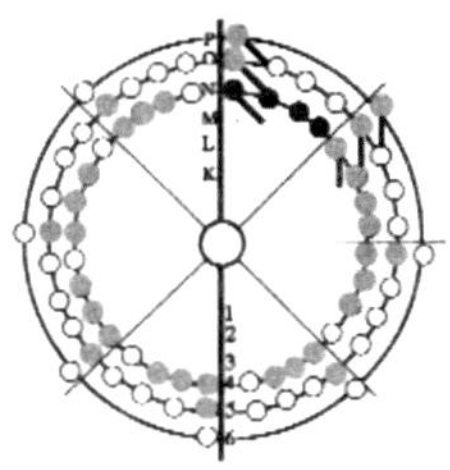

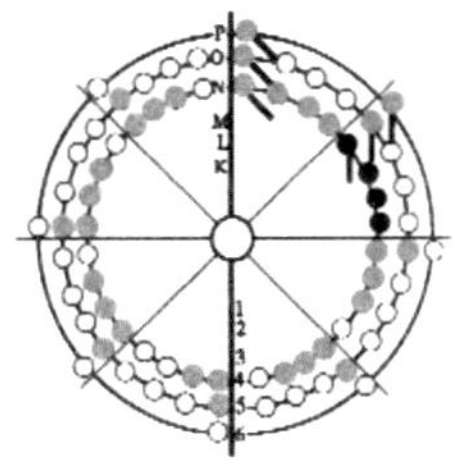

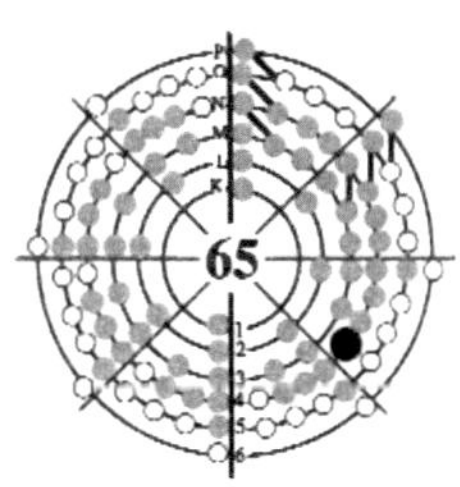

65. terbium
2-8-18-27-8-2

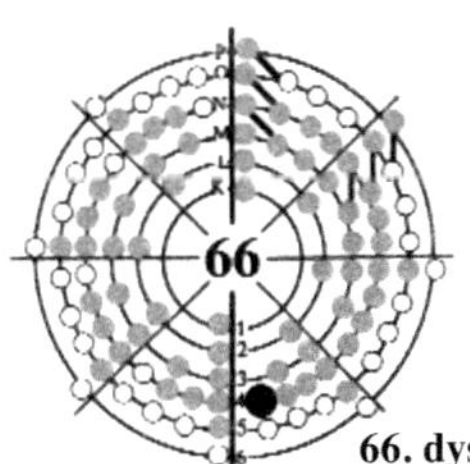

66. dysprosium
2-8-18-28-8-2

Eight Electrons Build on 5th Inner Shell and 6th Outer Shell

STEP 12

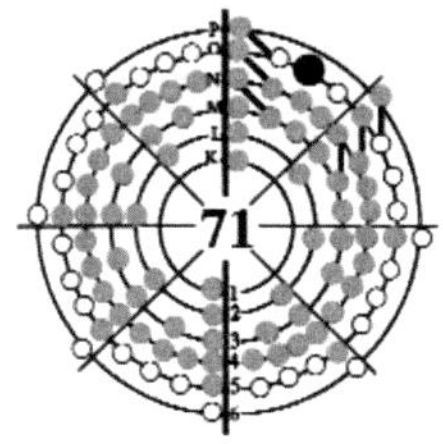

71. lutetium
2-8-18-32-9-2

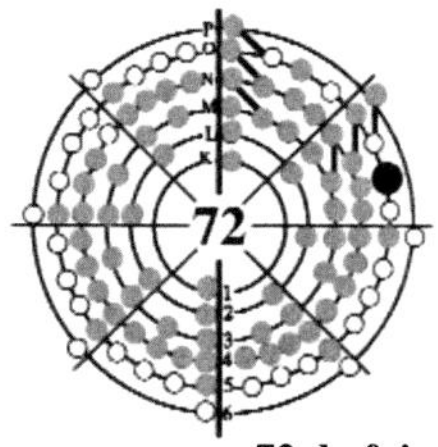

72. hafnium
2-8-18-32-10-2

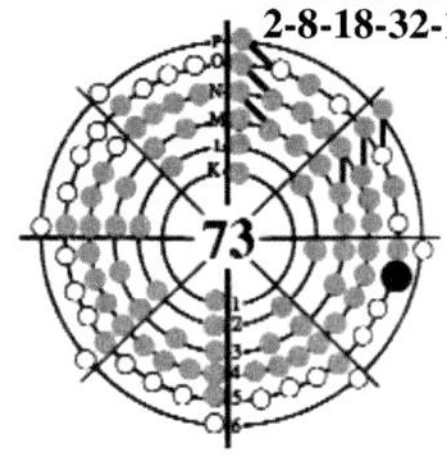

73. tantalum
2-8-18-32-11-2

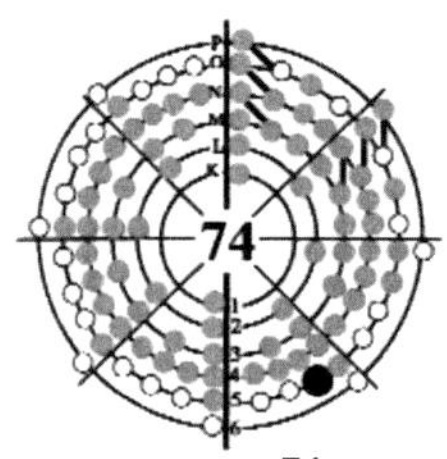

74. tungsten
2-8-18-32-12-2

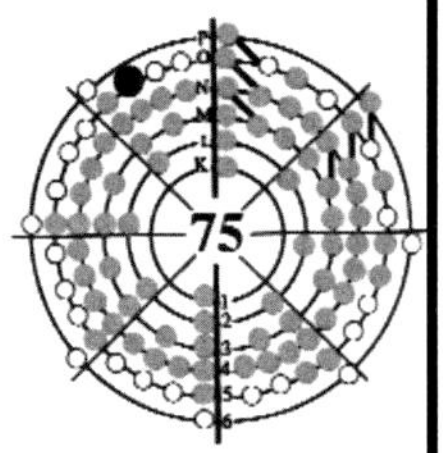

75. rhenium
2-8-18-32-13-2

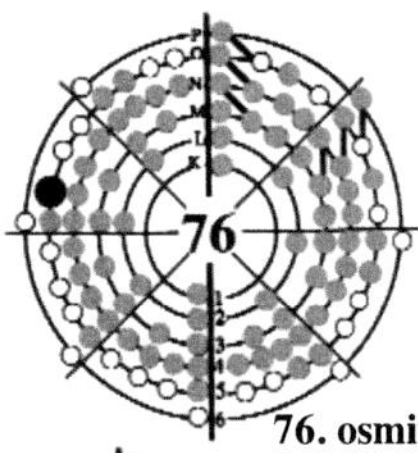

76. osmium
2-8-18-32-14-2

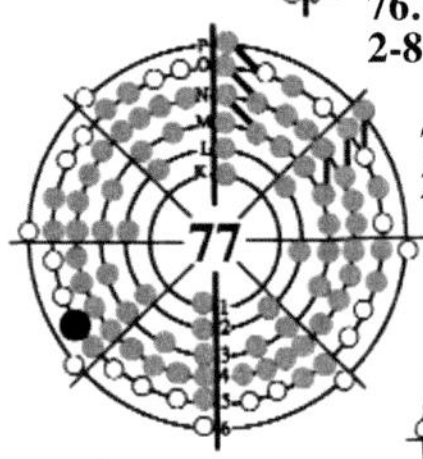

77. iridium
2-8-18-32-15-2

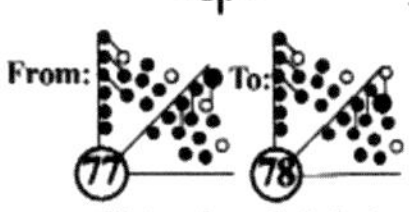

(Note enlarged circles)

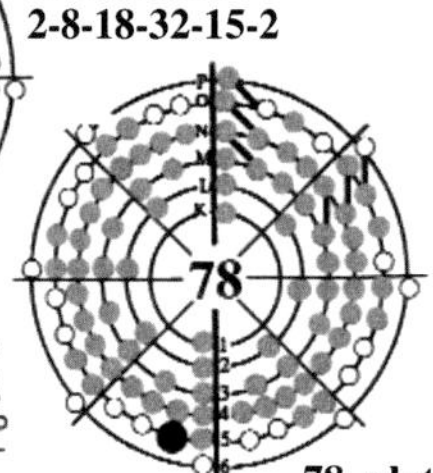

78. platinum
2-8-18-32-17-2

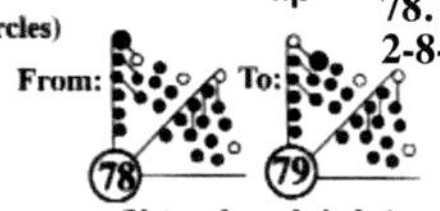

(Note enlarged circles)

STEP 13

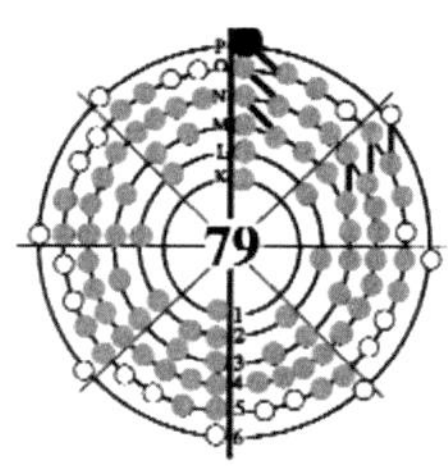

79. gold
2-8-18-32-18-1

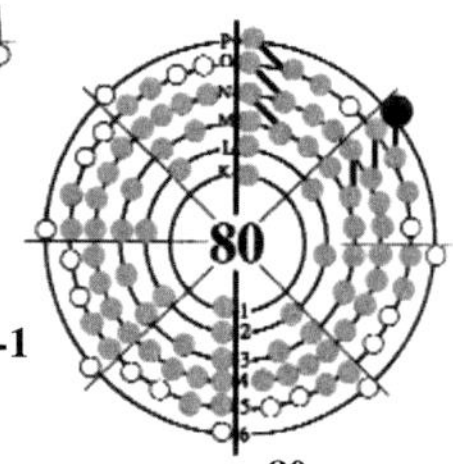

80. mercury
2-8-18-32-18-2

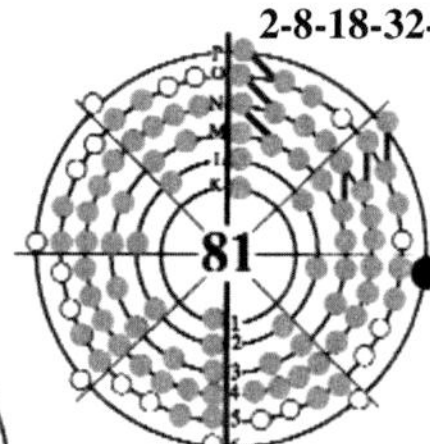

81. thallium
2-8-18-32-18-3

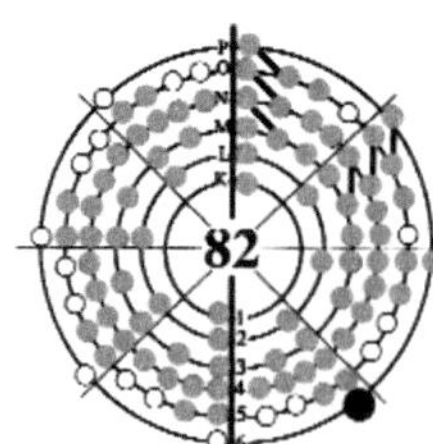

82. lead
2-8-18-32-18-4

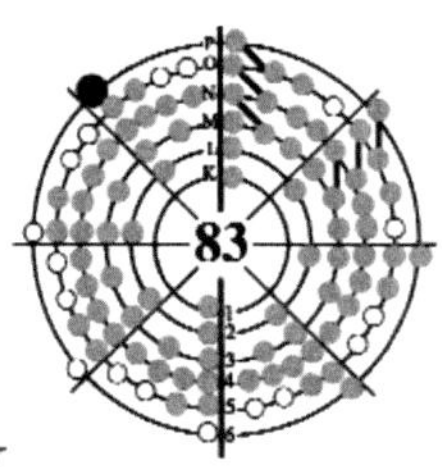

83. bismuth
2-8-18-32-18-5

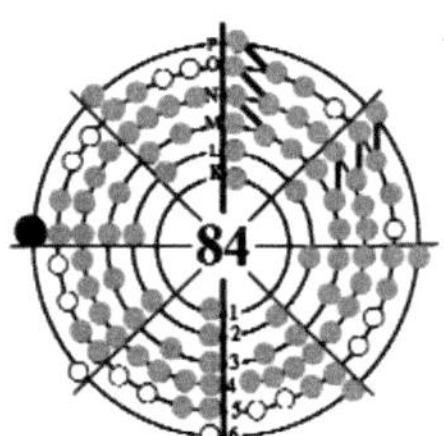

84. polonium
2-8-18-32-18-6

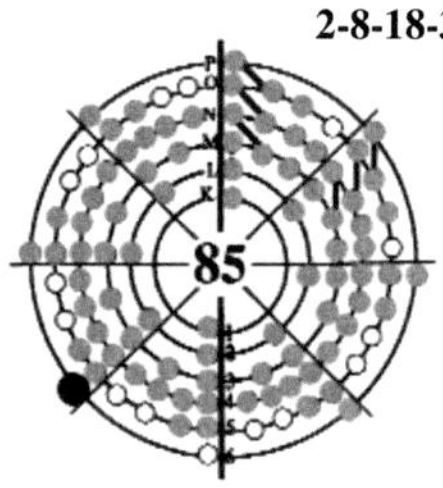

85. astatine
2-8-18-32-18-7

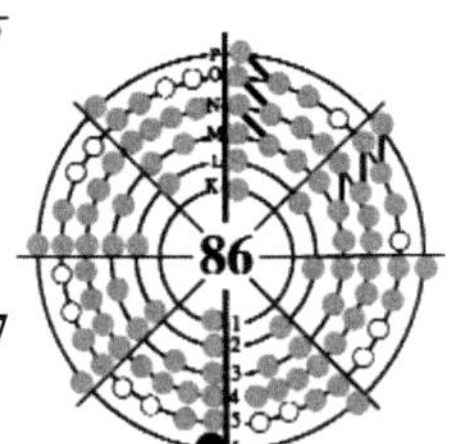

86. radon
2-8-18-32-18-8

Two Electrons Build on 7th Outer Shell – Four Electrons on 5th Inner Shell

STEP 14

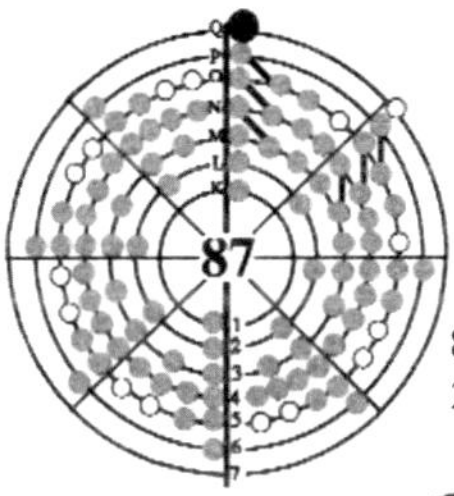

87. francium
2-8-18-32-18-8-1

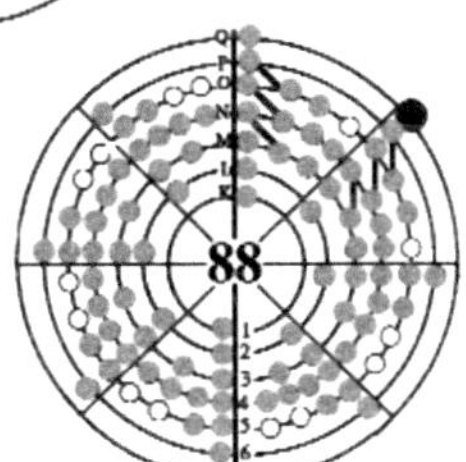

88. radium
2-8-18-32-18-8-2

Step 14. Although #87 and #88 are in "slider positions," they are not considered "sliders" because they do not drop into previous shells after setting the valence for transitionals #89 to #92. This is because the natural elements stop at #92. Should this pattern of expansion, yet contraction, continue to the man-made elements (and beyond) it would keep replicating itself.

Step 15. After #87 and #88 are placed on the 7th shell, add elements #89, #90, #91, and #92 to the inner space on shell 5. Since uranium, #92, is the last of the natural elements, Bohr's Chart stops here.

STEP 15

89. actinium
2-8-18-32-19-8-1

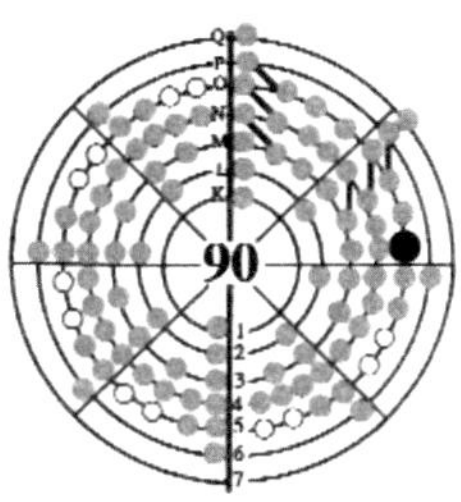

90. thorium
2-8-18-32-20-8-1

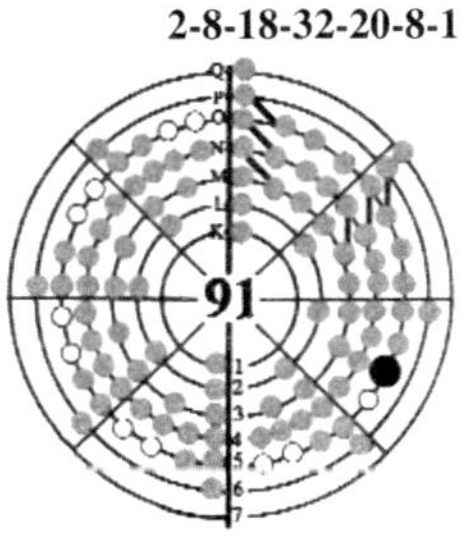

91. protactinium
2-8-18-32-21-8-1

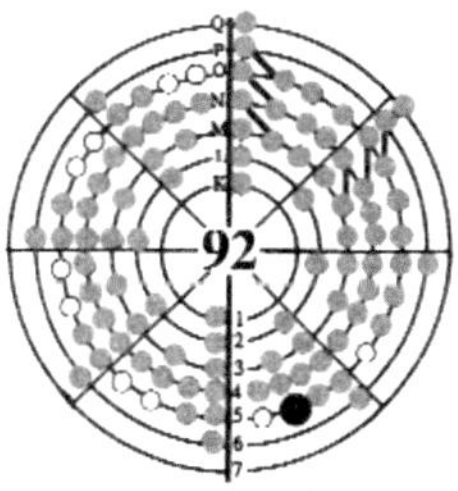

92. uranium
2-8-18-32-22-8-1

Congratulations! You have just made a model of an atom for each element. This may have taken some effort on your part, but remember, science has been searching a long time for the very information you have been given. You now know what the atoms of each element look like and how the electrons build one upon the other. You also have a knowledge of "sliders" that scientists have never understood. Let us now see how this model further clarifies already known scientific knowledge by giving us visual pictures that match the verbal descriptions.

VISUALIZING THE EXACTING ORDER WITHIN THE PAWLOWSKI ATOM

Refutes Electron Probability Distribution Within the Atom

The Pawlowski Atom effectively refutes the general randomness that science has placed on the positioning of electrons. Scientists speak of the probability or relative chance of finding an electron at a given location. The following illustration shows the weakness in this reasoning. Imagine taking 92 distinguishing electrons, throwing them up in the air, and then watching them land. How could uncertainty have positioned all eight families in their proper positions, along eight rays that are found in their proper energy fields?

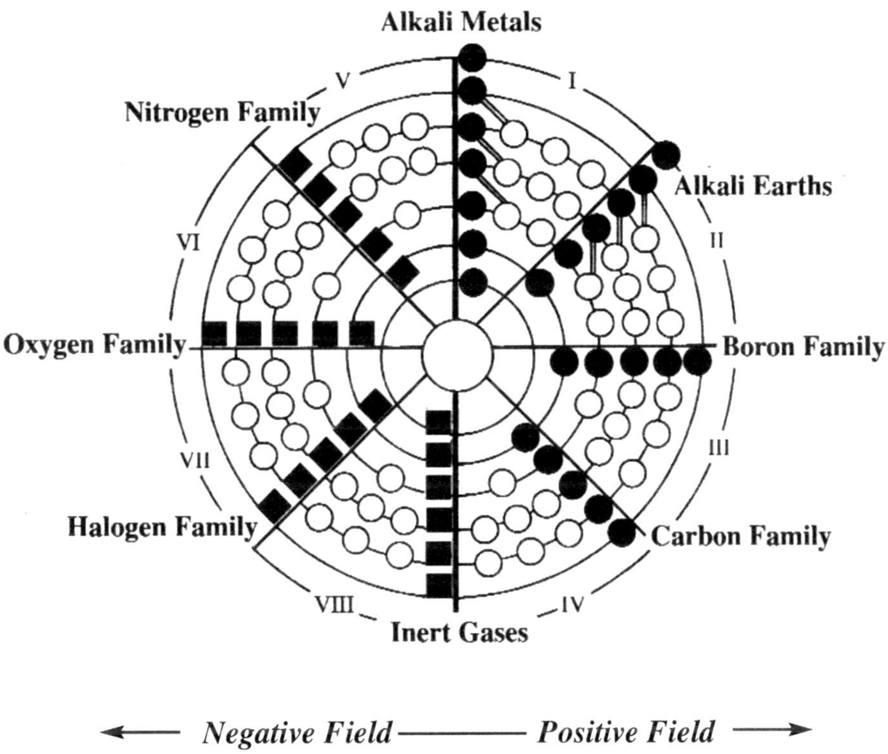

The circles along the rays represent the distinguishing electrons and elements in the positive energy field, while the squares along the rays represent the distinguishing electrons and elements in the negative energy field.

The Pawlowski Atom Provides Symmetry and Balance

There have been attempts at illustrating the simpler elements in the past, but a complete visual description of the atoms has never been successful because of the need for a definite positioning of the electrons. Since we now have a visual model, look at the two gold atoms (#79) below. Each has the same number of electrons, but the new model on the right, has symmetry and balance because of this exact positioning.

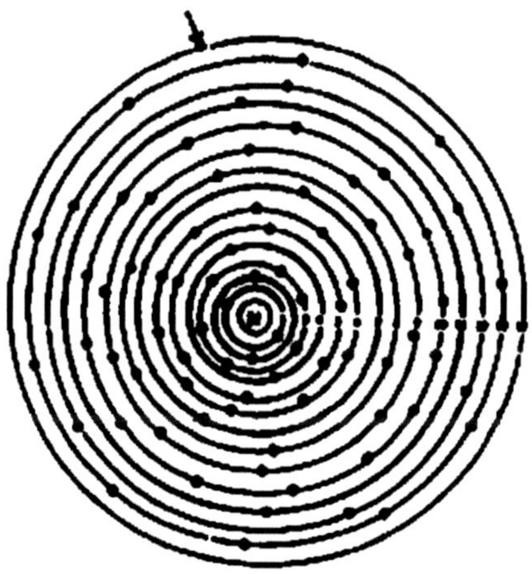

Standard Model of Gold

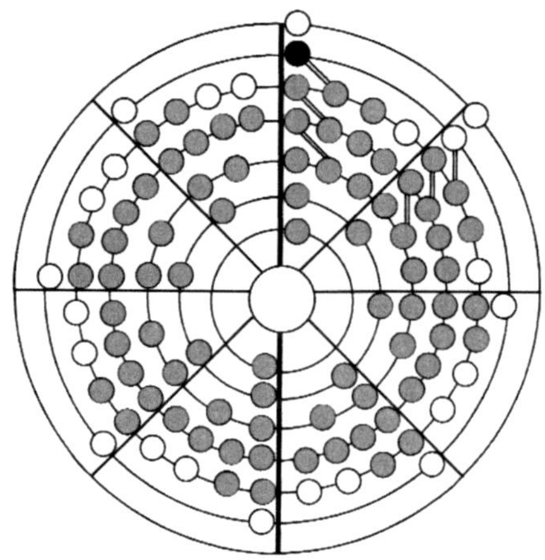

Pawlowski Model of Gold

The Verbal and Visual Descriptions Match

In Helen Pawlowski's book, *The Visualization of the Atom*, (pg. 72), we find an example of why her model is so effective as a teaching device. She says: "The atomic theory of the atom is based on the quantum numbers as set forth in the quantum theory. The entire theory is mathematical with some verbal descriptions to tie everything together. The circular periodic table/model of the atom now provides a visual description that fits into the mathematical theory of the atom, as well as much of the verbal descriptions. For example, the following quote concerning the 'inside shell structure' of several atoms was taken from the *Encyclopedia Americana*, (Volume 2, pg. 517 1964) under the subtitle of 'Shell Structure of the Atom:'"

> We observe...that beyond argon (Ar), whose M-shell is only partially occupied with 10 vacancies left, electrons begin to land in the N-shell, so that the M-shell is not completed until we reach the atom of copper (Cu), which contains 29 electrons. Similarly, the filling of the O-shell begins when the N-shell contains only 8 out of 32 electrons, and not until the P-shell has begun to fill up in ytterbium (yb), is the N shell completed."

Let us run through this again. This time the Pawlowski Model will be placed next to the verbal descriptions. By visually seeing the process, the mind can then better comprehend what is happening within the atom:

We observe…that beyond Argon (Ar), whose M-shell is only partially occupied with 10 vacancies left, (count the vacancies yourself)

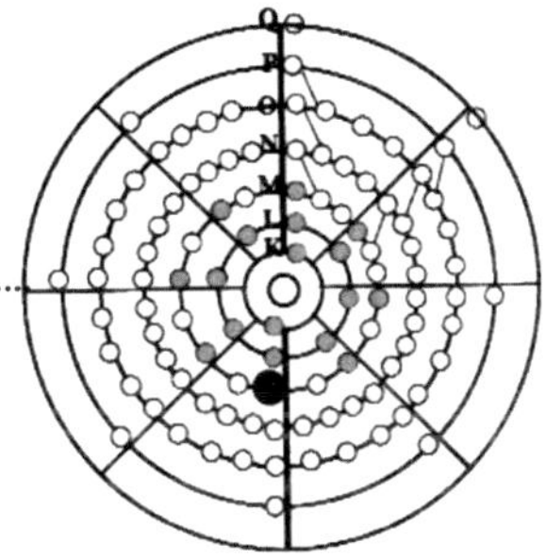

18 argon
2-8-**8**

electrons begin to land in the N-shell.

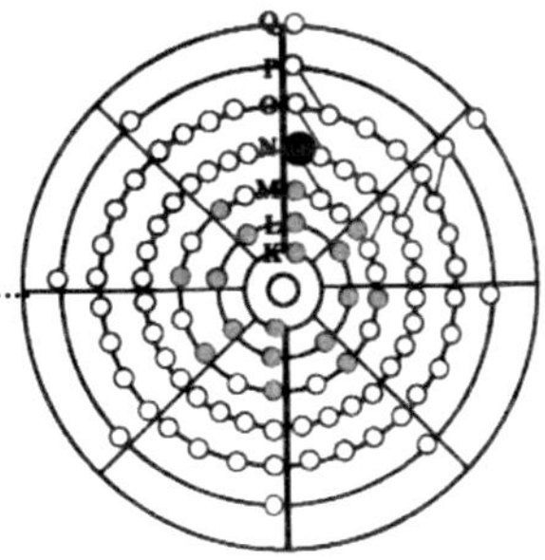

19 potassium
2-8-8-1

so that the M-shell is not completed until we reach the atom of copper (Cu), which contains 29 electrons,

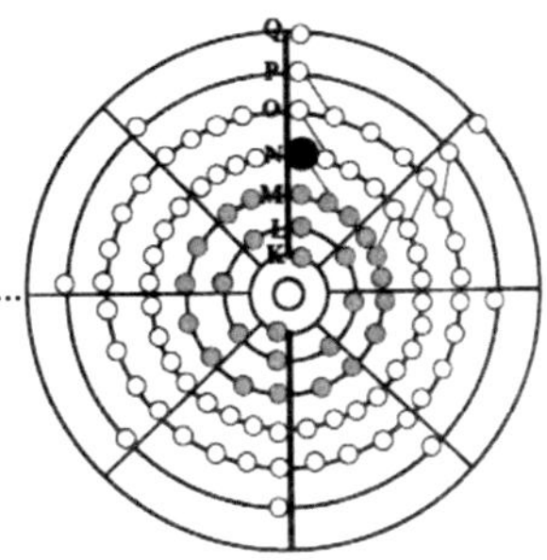

29. copper
2-8-18-**1**

Similarly, the filling of the O-shell begins when the N-shell contains only 8 out of 32 electrons,

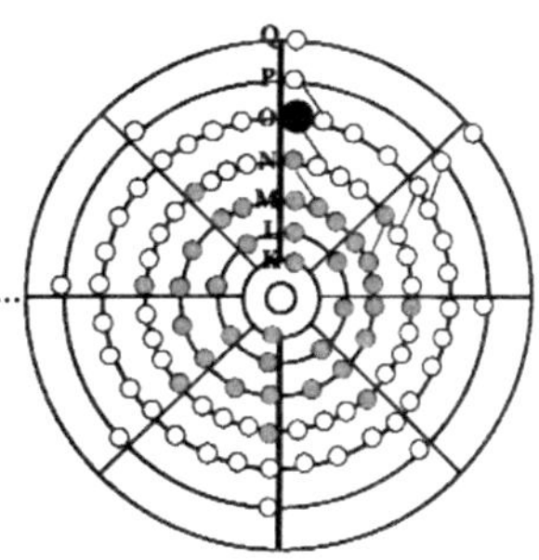

37 rubidium
2-8-18-8-1

and not until the P-shell has begun to fill up in ytterbium (yb) is the N-shell completed

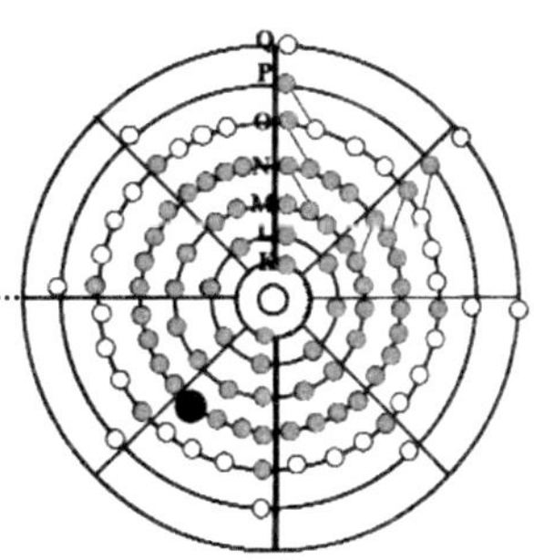

70 ytterbium
2-8-18-**32**-8-2

Visual Picture of Bonding

The Pawlowski Atom makes the bonding process more clear by visually showing how two atoms bond like a lock and a key and how the basic process of acting and being "acted upon" takes place between positive and negative elements. Below we see the right-sided positive element sodium #11 is giving its electron to the left-sided negative element, chlorine #17. The right-sided element does the acting while the left-sided element is acted upon.

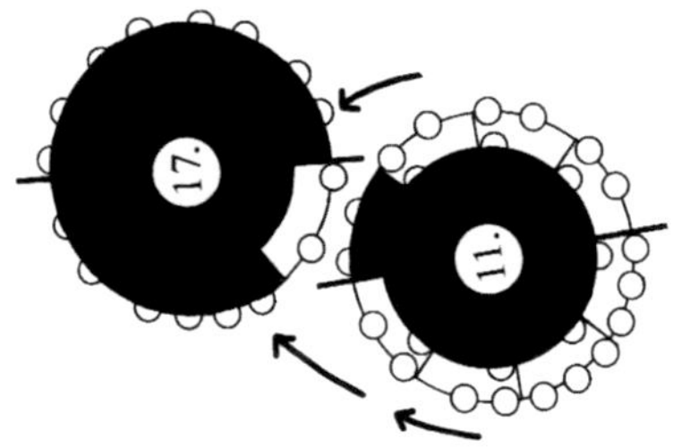

Polarity Barrier Establishes Type of Element

The Pawlowski Atom explains why three consecutive representative elements (such as #9 fluorine, #10 neon, and #11 sodium) can be negative, neutral, and positive, yet the only difference between the three is one electron per element. It is the placement of the electron in relation to the polarity barrier that establishes whether an element is positive, negative, or neutral. The transitionals are different. Their positiveness comes from the two slider electrons on the right, outer most shell.

#9 Fluorine / Negative — #10 Neon / Neutral — #11 Sodium / Positive

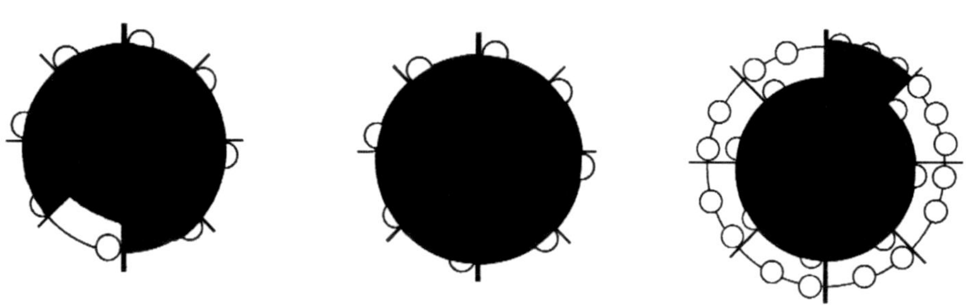

Spin Up – Spin Down

Science knows which electrons are spinning up and which are spinning down. When our brother Paul Williams superimposed this information on the Pawlowski Atom, we were surprised to see the electrons working together and generating power on an even deeper, more complicated level within the atom than the alternating of the two positive and negative energy fields. It broke the quadrants down into octets that manifest an opposition that would be impossible to duplicate if it were left up to probability/chance. It showed the spin of the electrons in each octet working in opposition to its opposite octet. With the electrons in one octet spinning up (the light shadings) and the electrons in the octet directly opposite spinning down (dark shadings), it gives evidence that the alternating of opposite energy is what causes the electrons to spin. This model also provides evidence that the electrons remain seated where they originally dropped, and that the figure eight electron path is indeed the path that delivered the electrons, since the "spin up" "spin down" octets also change directions at the polarity barrier. If just one electron was out of place, this exacting order within the atom would not be possible. The following quote supports Helen's illustration of spin:

> Suppose that we have what the physicists call a two party system of zero spin. This means that the spin of each particle in the system cancels the other. If one of the particles—has a spin up, the other particle will have a spin down. If the first particle has a spin right, the second particle has a spin left. No matter how the particles are oriented, their spins are always equal and opposite. (*The Dancing Wu Li Masters*, by Gary Zokov, Page 283)

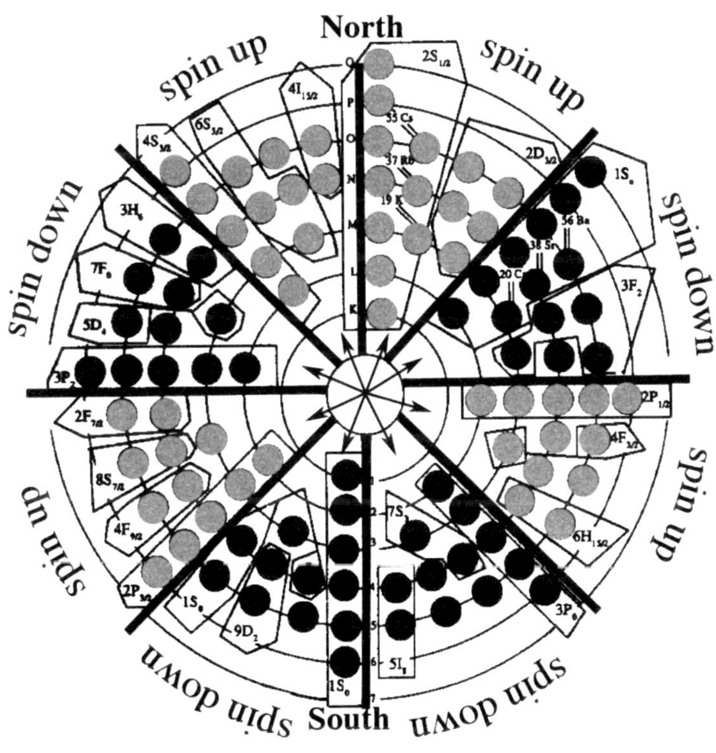

The Anomalous Zeeman Affect

The Pawlowski Atom visually illustrates how the "sliders" in the upper right quadrant create an unusual shifting of electrons in the alkali metals and alkali earths. These unusual movements are essential to the placement of the electrons. A man by the name of Pieter Zeeman first recognized these anomalies; thus they are called the Anomalous Zeeman Affect. Another scientist by the name of Wolfgang Pauli spent time and energy investigating the action of an atom in a magnetic field and its ensuing line spectra. His studies indicated there were some "loose parts" within the atom. Pauli concluded that the Anomalous Zeeman effect was a phenomenon not general to the atom as a whole, but were restricted to the alkali metals and alkali earths of group I and II. *See arrows.*

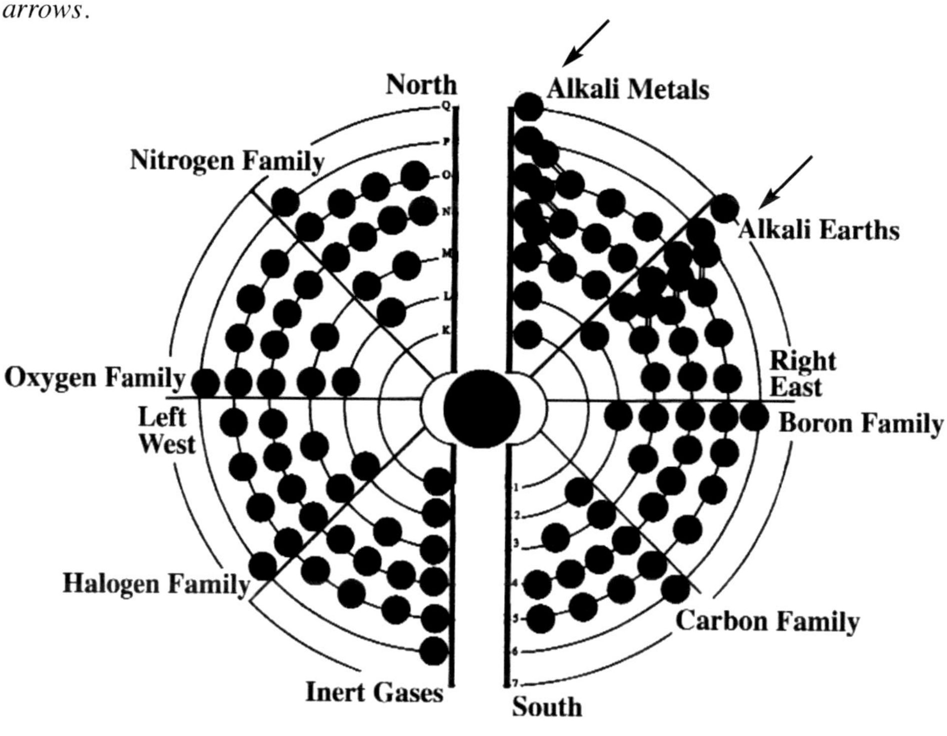

Balance and Equilibrium Within the Elements

The above illustration also shows how the six extra "slider electrons" in the upper right quadrant create a strong force *on the right side of the polarity* that keeps the momentum and thrust of the atom moving forward. The electrons throughout the rest of the atom are almost equally distributed and balanced.

Helen Pawlowski likened the balance and equilibrium we see in the atom to the spinning of clothes in an automatic washer. Should too many be on one side of the tub, it would wobble erratically from side to side. The clothes must be evenly distributed for stability to be established. It is the same within matter. The electrons must be evenly distributed in two energy fields in order for them to oscillate back and forth in generating power.

Why are Negative Representative Elements on the Left Side so Volatile?

Could the elements on the left side of the atom be spring loaded? Since the positive field has nothing that restricts its energy, it would seem that as the elements form on the opposite left, negative rays, matter would be brought into line by the magnetic attraction of the positive and negative elements making contact. Since negative elements cannot give, they hold the line by putting on the brakes. This pulling back creates a tension that positive elements do not have. It is interesting that the most volatile of the negative elements are those seated near the center, on the left rays, while those further out, are less volatile.

Once the pressure coming from the spring-loaded elements on the first three rays is released, matter becomes balanced and whole, thus creating an inert gas.

The following shows the volatility of the negative elements:

> ***Nitrogen*** *#7*, is the first negative element to be formed. It is on the left side of the first shell and the first of five on the first ray. It is used for explosives, fertilizers, and dyes. Next ***Phosphorus*** *#15*, is poisonous, inflammable, and luminous in the dark, and is used for matches and fertilizers. ***Arsenic*** *#33*, is next and is volatile when heated and forms poisonous compounds. Then ***Antimony*** *#51*, is a brittle, lustrous metallic element used chiefly in alloys and medicine. Next ***Bismuth*** *#83*, is a brittle metallic, having compounds used in medicine.
>
> ***Oxygen*** *#8*, is on the negative side; first shell, second ray. It is a colorless, odorless, gaseous element. ***Sulfur*** *#16* is next. It is used in making gunpowder and matches. Then ***Selenium*** *#34*, is a nonmetallic element chemically resembling Sulfur and Tellurium. ***Tellurium*** *#52* is next, a rare silver white element resembling sulfur in its chemical properties. It is little used because of its poisonous nature. Then ***Polonium*** *#84*, is a radioactive element discovered by the Curie's Radium.
>
> ***Fluorine*** *#9*, is on the negative side; first shell, third ray and is just before the inert gases. It is a nonmetallic element, a pale yellow corrosive gas. Next is ***Chlorine*** *#17*, a gaseous element in common salt, incombustible and highly irritating to the organs of respiration. Then ***Bromine*** *#35*, is a dark red fuming liquid, resembling chlorine and iodine in chemical properties. ***Iodine*** *#53*, is next and is used in medicine as an antiseptic. Then ***Astatine*** *#85*, is a highly unstable radioactive element used in medicine as a radioactive tracer.

S, P, D, F Numbers

How do the s, p, d, f numbers fit into the Pawlowski Atom? Helen explained it to me in this way, "Imagine you are looking through a powerful microscope, directly *into* the center of the atom from the right, east side. Because the light emanating from some electrons appear brighter than others, the letter s, p, d, f, were given to distinguish their positions by the light they impart. The light coming from the "s" electrons are sharp, the "p" electrons are principle (primary), the "d" electrons are diffuse (dim), and the "f" electrons are fundamental (faint, deep). When "slider electrons" are on the outer most shell, they are the sharpest of all because they are the nearest to where you are looking. The electrons in the inner shells are the least sharp because they are the farthest away. The "s" and "p" electrons are always on an outer most shell along one of the eight rays. The "d" electrons are always on the second from the last shell, while the "f" electrons are always on the third from the last shell. The "s" and "p" electrons build on the outer shells, the "d" and "f"on the inner shells. This is another example how the Pawlowski Atom makes it easier to visualize already known scientific facts.

J.S. Bells Theory

The following quotes from *The Dancing Wu Li Masters*, by Gary Zukov, points to a recognition by some in the scientific community that there is a different state, much like you would find in an energy field that embodies religion:

> Bell's theorem could be the Trojan horse in the physicists camp; first, because it proves that quantum theory requires connection that appear to resemble telepathic communication, and second, because it provides the mathematical framework through which serious physicists could find themselves discussing types of phenomena which, ironically, they do not believe exist. (Pg. 298)
>
> One of the implications of Bell's Theorem is that, at a deep and fundamental level, the separate parts of the universe are connected in an intimate and immediate way. (Pg. 282)
>
> Professor G. F. Chew, Chairman of the physics department at Berkeley, remarked, in reference to a theory of particle physics: "Our current struggle (with certain aspects of advanced physics) may thus be only a foretaste of a completely new form of human intellectual endeavor, one that will not only lie outside physics but will not even be described as scientific." (Pg. 314)

Testimonials

Doctor Dell K. Allen, who for many years was a Professor of Manufacturing Technology at Brigham Young University, became acquainted with Helen Pawlowski and her model of the atom while teaching at the school. After becoming familiar with her chemistry material, he met with her brother, Paul Williams, to discuss the extensive physics research he had done concerning the correctness of the Pawlowski Atom. Since Dr. Allen already understood the approach Helen had taken, he could readily see how the physics part fits in with the chemistry part in explaining atomic structure. After their visit, Dr. Allen wrote Paul a letter that contained the following quotes:

> It is my opinion that the work you and Helen have done is very innovative and creative and best of all, it is mutually supportive of one another's work. The Circular Periodic Table that Helen has been working on for so long seems to provide a simple and useful framework for explaining important relationships in atomic structure, and the work that you have done provides an outstanding scientific support for applications of this model in all facets of science. *Together your work is revolutionary and exciting*.
>
> ...This new approach you have developed is so overwhelming in its scope and potential that it can provide the basis of work for hundreds of researchers who accept these new concepts.
>
> ...*I believe your work represents the grand unifying approach to the universe for which so many have been searching*. (emphasis added)

Thanks to Dr. Allen, Helen was able to set up her book, *The Visualization of the Atom*, in a professional manner. When he first saw her work, her illustrations were all drawn out by hand. Obviously she needed help and Dr. Allen was just the person she needed. Because of a previous grant, he was already established in a computer lab at Brigham Young University and teaching students how to put scientific material on computers. Being impressed with her work, he wanted to see that it was published and disseminated correctly. He and his students, therefore, spent hours in the lab teaching Helen how to construct the illustrations in their present form. I cannot thank him enough for the important role he played at that beginning time.

Many others throughout the years have studied Mrs. Pawlowski's, *Visualization of the Atom*, and found it to be correct. One of the first persons to be approached to see if he would critique Helen's material was Dr. Thomas Barrett, an agronomy teacher with a background in chemistry, who had just retired after teaching 37 years at Arizona State University. After studying her material he said, "I can find no fault with the work." Dr. James E. McComb, my Rheumatologist, also said, after studying her book, "Do you realize that every chemistry book in the world will have to be changed." It is my belief that not only chemistry and physics students will greatly benefit from Helen's efforts, but also the world at large. By creating a visual picture of how the electrons are organized, this pattern of creation presents a new and exciting alternative to conventional thinking. Besides giving a more thorough knowledge on the inner workings of the atom and a deeper understanding of how matter throughout the universe has been structured, it also gives an all encompassing picture of the role that opposites play in the world around us. Truly Rick Gore was right when he said, "In the structure and behavior of the smallest things, lies the explanation of all matter."

The pictures that I have attempted to paint within the mind are just part of the equation, for only the heart can concede that God created such exacting order and that the repeated pattern of opposites did not happen by chance.

***

More in-depth scientific information on Helen Pawlowski's work can be found in her chemistry book, *The Visualization of the Atom*. It is distributed by the Pawlowski Family Trust, P. O. Box 1483, Riverton, Utah 84065, kubricky@earthlink.net. Also, Paul A. Williams has gathered a vast amount of evidence to show how the laws of physics dovetail into the Pawlowski Atom. He is in the process of writing a book called, *Evidences and Ramifications of the Circular Periodic Table/Model of the Atom*. Since his material is so expansive and far-reaching, much of it has to be verbally transmitted. For information contact, Paulallenwilliams@yahoo.com.

OPPOSITION IN ALL THINGS

The existence of two opposite energy fields is the most unusual discovery to come from the Pawlowski Atom. The opposition factor between the nucleus and surrounding electrons has long been recognized, however, the right-side–left-side dipole feature has not. By placing the electrons in exacting positions in a circular configuration; first to the right, then to the left of a north/south polarity barrier and by taking into account previously unaccounted for movements in the "slider" area, this aspect is clarified. This separation of matter into positive and negative energy fields should not be too surprising, however, as one need only to study the physical sciences to realize there is a rule of opposites at the very core of nature. Albert Einstein said matter must have both a positive and a negative component. The Pawlowski Atom has shown this to be true.

Life has so many dissimilar, yet complementary opposites that we could ask, "How did these opposites form? Was it by evolution or probability/chance or are things such as plus-minus, east-west, right-left, alkaline-acid, day-night, treble-bass, attraction-repulsion, spin up-spin down, etc. simply manifestations of what is happening within the atom?" To me the answer is obvious. Since everything is made up of either positive or negative atoms/elements, this opposite structure will naturally expand throughout everything in the universe.

Magnetism and Balance: The Two Forces That Hold Things Together

As a young girl, I often wondered, when looking into the night sky, why the stars in the heavens did not fall into space? And why did the earth not do the same? And when looking at pictures of heavy, cumbersome planets, I would wonder how such enormous things could stay in place. I now believe it is because an elaborate system of positive and negative opposites have been set up in the heavens that creates a magnetism and balance between planets.

On the next five pages are illustrations from Helen Pawlowski's unpublished book, *Mankind: A Compound in One*. They give a clearer picture of how she felt the *law of opposites* works. They illustrate how the magnetic attraction between positive and negative energy is what pulls matter back into line and that balance is not an external thing that lie outside matter, but is embodied within matter. They show, in a sim plified way, how the law of opposites creates balance and wholeness. Because of this opposite structure within the atoms, everything points to a universe that seeks balance and equilibrium. Even the plants, animals, and humans that reside here on earth are composed of two parts that seek balance and equilibrium.

Opposite Energies in the Plant Kingdom

1. All true complementary modes are based upon two components of equal but opposite charge, i. e., positive and negative energy.

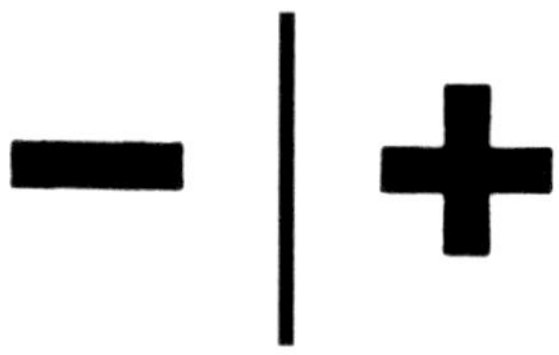

2. All plant resources are ultimately derived from the sun (+) and the earth (–). Although the earth contains both positive and negative elements, at its center is a magnetic core. Magnetic energy is always negative because it draws things to it. This energy is also known as gravity.

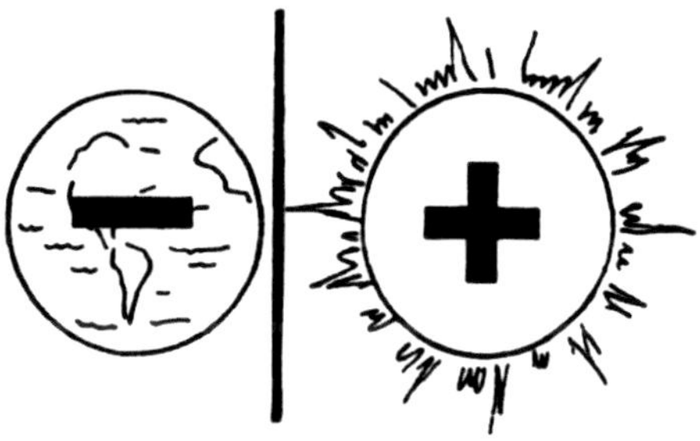

3. All parts of a plant are capable of differentiating between the energy of the sun (positive +) and the energy of the earth's magnetic core (negative –). Plants respond to these two types of energy by growing towards them. Growth at the top part of the plant is drawn towards the sun (phototropism) while growth at the root part is drawn towards the negative energy of the earth's magnetic core (geotropism).

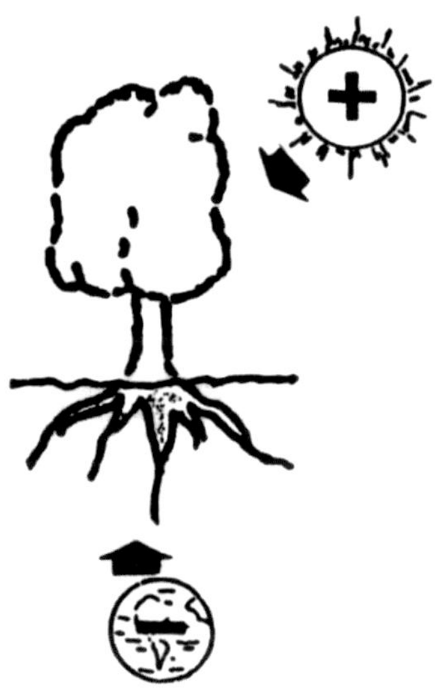

4. According to the law of opposites, two things that are polar opposites are attracted to each other. Two things that are the same are repelled. (This applies only to physical things).

5. Therefore if the sun is positive, then somewhere within the part of the plant above ground is a source of negative energy or ionized negative elements.

6. On the other hand, because the earth's magnetic core is negative, it attracts positive energy or positive ionized elements in the part of the plant below ground.

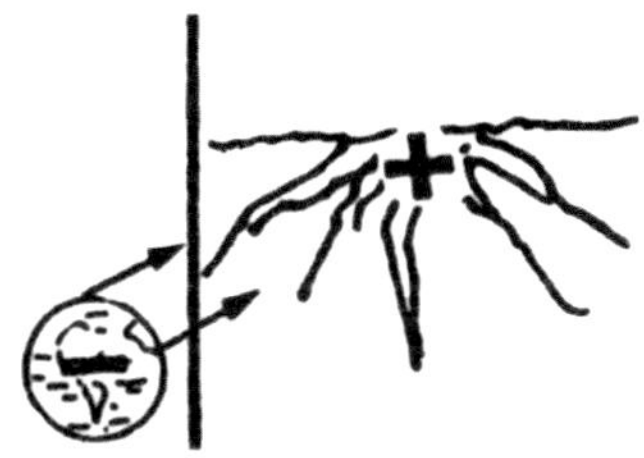

7. With the normal passing of time, the top part of a plant becomes so lofty that the ratio between the positive part (the roots) is no longer equal to the negative part above ground. It is the inequality between the two main components that eventually leads to the death of the tree. Pruning a mature tree is beneficial because it helps maintain a balance between the negative energies at the top of the tree and the positive energy in the root system. Bonsai, the Japanese art form of keeping a tree alive indefinitely, is based upon an equal top and root system.

Opposites have everything to do with balance between two parts—the positive and the negative. As long as the two parts remain balanced, life is sustained. Although opposites start with small things then expand into large things, they are the very foundation of our existence. Without them there would no be life.

Opposite Energies in the Animal (Mankind) Kingdom

Since a compound is something composed of two or more parts, or something that cannot stand alone, are we humans considered a "compound in one?" *If the Pawlowski Model of the Atom is true*, and there is a pattern of opposites governing the smallest atom, *then everything, even our bodies, must reflect this duel structure*. How else could elements be animated, except this opposite structure be present, and how else could maximum power, energy, magnetism, and light be produced, unless two bodies work together as one?

Now I don't pretend to be a scientist nor do I pretend to have all the facts, but I do feel the Pawlowski atom makes it possible to visualize the opposite nature of things. It gives evidence that we do have a positive spirit body that wants to act and a negative physical body that wants to be acted upon and that together they have been made to work as one.

Perhaps this next illustration will give a broader view of this opposite structure. The human body, on the basic level, is composed of 11 main elements plus minute amounts of trace minerals. Measuring the ash and gases that are left after cremating a human body is how the ratios were arrived at. Although there are slight variations between various authorities, the following shows an approximate percentage of each of these elements and whether the element is a positive or negative.

Composition of the Physical Body

Elements	Negative	Positive
10.5 % Carbon		10.5 %
10.0 % Hydrogen		10.0 %
76.0 % Oxygen	76.0%	
2.5 % Nitrogen	2.5%	
0.05% Sodium		0.05%
0.02% Magnesium		0.02%
0.3 % Potassium		0.3 %
0.02% Calcium		0.02%
0.3 % Phosphorus		0.3 %
0.2 % Sulfur	0.2%	
0.1 % Chlorine	0.1%	
Sub Total: 99.99% *	78.80% *	21.19% *
0.1% Iron and other trace elements		
Total: 100%		

*These numbers will now be referred to as 80% negative and 20% positive.

This chart on the composition of the physical body indicates that human bodies are made of 80% negative elements and 20% positive elements. But if the physical body is made mainly of negative atoms/elements, how did we become a living soul with such a lopsided ratio within our bodies? Shouldn't another entity, made of 80% positive and 20% negative atoms/elements, be added to make it whole?

To my thinking, this would have to be so. Therefore, I will proceed on the premise that if balance and equilibrium are to be established and a magnetic attraction set up, each living soul must be composed of two bodies that have opposite positive-negative energies. As shown in the following illustration, one body would be composed of mainly positive, giving-type elements that want to "act," while the other body would be composed of mainly negative, receiving-type elements that want to be "acted upon."

The Miracle of Life

A Physical Body
made of 80% negative
and 20% positive element

and

A Spirit Body
made of 20% negative
and 80% positive element

Coming Together
to
Create a Living Soul

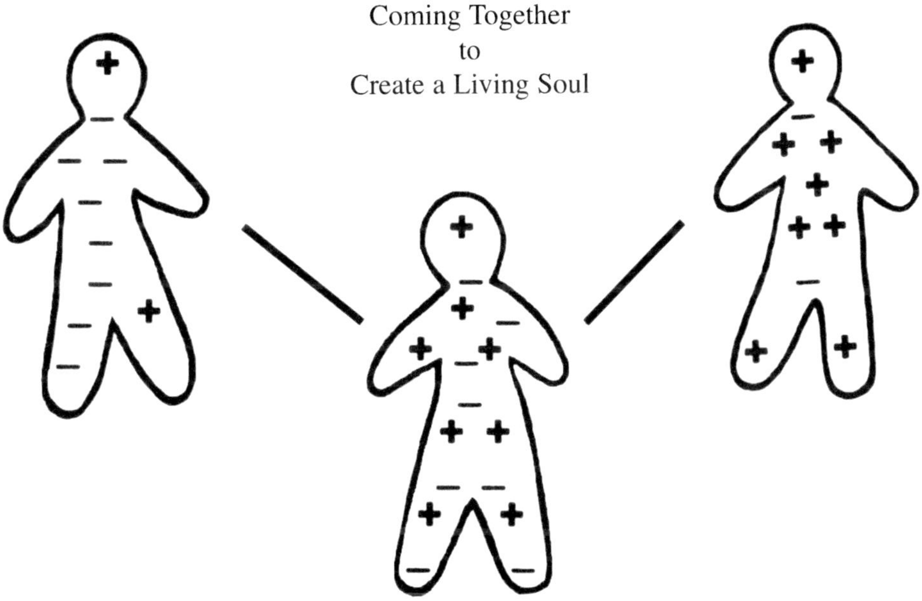

When speaking of elements from which we are made, spirit elements will now be referred to as "positive" and physical elements as "negative." This may not be in strict scientific terms, but the preponderance of positive is always on the right while the preponderance of negative is always on the left.

To Act and Be Acted Upon: To Give and To Receive

What are the characteristics of the spirit and physical body? The spirit body is a giving entity, while the physical body is a receiving entity. The spirit body "acts," the physical body is "acted upon." How do we know this? These next silhouettes show how positive elements have electrons projecting out to give or share, while negative elements have empty spaces to be filled. The predominant elements of each body will display its dominant characteristics. Therefore, when the spirit body shares its electrons with the physical body, the two bodies become balanced and whole.

The Giving and Receiving Attributes of the Positive and Negative Elements

<u>Negative</u> receive electrons

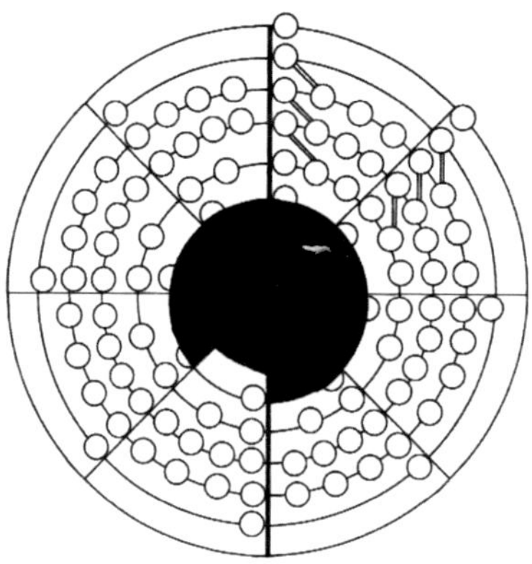

<u>Positive</u> give and share electrons

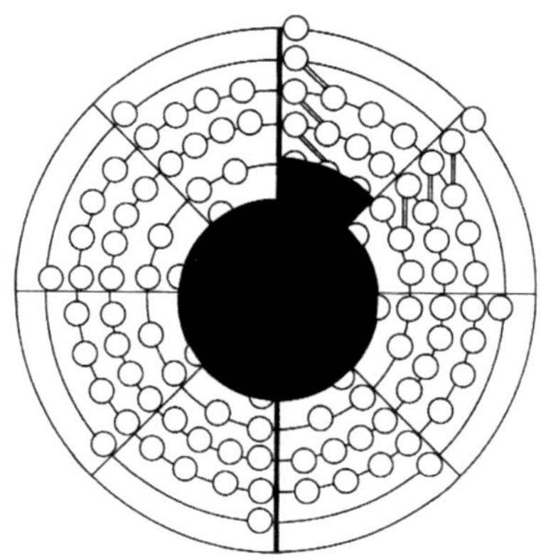

This next illustration reveals the importance of giving and sharing. By having the positive element, sodium #11, share its electron with the negative element chlorine #17; salt, NaCl is created. This gives evidence that when positive elements give and negative elements receive, the two are added upon and become greater than the one.

#11 Sodium

#17 Chlorine

Another example of matter building by a process of giving and receiving, consider the composition of water, which is $H^{2}0$. Hydrogen (#1), a positive element, has one electron it can share while oxygen (#8), a negative element, has two empty spaces for other electrons. When two hydrogen atoms bond with an oxygen atom that has two empty spaces, water is formed. Hydrogen by itself is not too important, but when it combines to make water, it becomes part of the substance that covers a large part of the earth. It can then float in the clouds, fill the oceans, rivers and streams, as well as become a necessary part of every living thing. Hydrogen and oxygen become balanced and whole when they join together, with one "giving" and the other "receiving." They complement one another by filling the other's need. Since complement means, "that which completes or makes perfect," by coming together as positive and negative complementary opposites, they are able to become perfect and complete.

When hydrogen and oxygen stand separate and alone, their natural inclination would be to go in opposite directions. But when standing side by side, yet opposite one another, they would magnetically attract because one is a positive element, the other is a negative element. The following illustrates this phenomenon:

> Positive: measured or proceeding in a direction of increase, progress, or onward motion.
>
> ***P o s i t i v e*** **-------------------------------->**
>
> Negative: involving denoting subtraction: minus, measured or proceeding in the opposite direction to that which is considered as positive.
>
> **<-----------------------------------*N e g a t i v e***
>
> Put the positive and negative together and they naturally attract and fill the other's need.
>
> **-----------------------> <-----------------------**

Elements R Us

The subject *Elements R Us* is derived from the popular store *Toys R Us*. Even though we tend to not look on elements as part of the equation of life; they are the building blocks of creation. Take them away and we have nothing, for all things are made up of elements. This is why it is so important to know something about them—it gives us a glimpse of who we really are. The above explanations will now be applied to the elements of our own body, for if positive and negative elements can unite to form compounds, why not unite to form a human body? Some may say this is a giant leap of faith to go from something as small as elements and atoms to something as large as man. But keep in mind that just as in building the atoms—when we developed and grew within the womb—each new part of our body was building and expanding upon a part that had previously not been there. Obviously, creation is a step-by-step process of being added upon.

At a given time during the process of creating a living soul, the positive spirit body and the negative physical body unite to become one body. To visualize two bodies working together in the same confining structure, imagine looking through a powerful microscope at the minute matter making up these two bodies. In the physical body you will see mainly negative elements that have holes within its structure that can be filled with electrons. In the spiritual body you will see mainly positive elements that have electrons jutting out that can be given away or shared. Then imagine the spirit body giving and sharing its electrons with the physical body. In this way, one body is able to give; the other filled and satisfied. The two bodies can then work together as one.

Now picture these two bodies not working together as one. Although each body is intertwined in its interaction with the other, it appears that whenever something goes wrong, the root cause always lies within the physical body. The reason for this assumption is that the physical body has the ability to take in and absorb harmful physical things that are detrimental to the two bodies, thus upsetting the balance between spiritual and physical energies. To better visualize this imbalance, picture in your mind the positive spiritual body and the negative physical body as an electrical unit that is generating power, magnetism, energy, and light. Should too many negative things enter this two-part system, undesirable consequences would emerge. The unit made up of two parts would then become like a flashlight with weakened batteries—its light and power would be diminished. As the negative continued to draw from the positive, darkness would eventually come and the balance between the two would be gone.

Add or Subtract – Plus or Minus

The Pawlowski Atom reveals that all organized matter is set into either positive (+) or negative (–) fields of energy. This gives evidence that the human body can be described in these same mathematical terms, for it too is made-up of two bodies with the same two energies. Because the spiritual body is made-up primarily of positive "giving type" elements, it has, by its very nature, a progressive type motion that moves it forward (+). This ability comes from the right, positive side of the atom having more electrons and activity than the left side. It would seem that as the spiritual body gives and shares its electrons, it would end up with less (–) positive energy, but instead, as the definition of positive implies, it proceeds in a direction of increase, therefore it ends up with more (+). It is like the old adage, "the more you give, the more you receive." This old axiom is a true principle originating within the elements, for the more we exercise the spirit side of our being by serving and helping others, the more positive (+) energy is created to keep the physical body filled and satisfied. Both bodies would then proceed in a direction of increase, progress, and growth.

What is the opposite of increase, progress, and growth? It is negatives moving in a reverse direction. Instead of increase, there is decrease. Because the physical body is made-up primarily of negative, "wanting to receive" elements, it might seem that the more negative energy this body receives, the more it would have. But this is not so, for should it receive more than is necessary, the more it would have to subtract or take away. Simple math illustrates this concept. Start with (–2) and add to this another (–2). The answer is then (– 4), a bigger negative. On a number line (–100) is less than (–50). Greater negatives grow in a decreasing direction so the more one gets, the less one actually has.

If increase and progress lead to success, the opposite leads to failure. Because negative receiving type elements proceed in a negative direction, when excessive negative (–) energy/atoms enter the body, they can only be satisfied if there are enough positive (+) energy/atoms to go around. This is why we should keep our physical wants and needs in proper balance. Things such as food, shelter, and clothing, along with our desire to procreate, do create needs and desires that must be met. But should the physical body have more than is needed, excess negative energy is generated.

It is clear then why things of the flesh do not satisfy. Accumulating more negative energy in the physical body than the giving spirit body can supply leaves more empty spaces to be filled and satisfied. Attempting to satisfy this emptiness by physical means then puts even more negatives wanting electrons within our bodies. Because our physical and spiritual bodies are designed to fill and satisfy the needs of each other, our lives must be balanced between our giving spiritual natures and our receiving physical natures.

Have I given the impression that positive spiritual things are a blessing while negative physical things are undesirable? Let me erase that idea once and for all, for the positive and negative are of equal importance since you can't have one without the other. The key is to keep them in balance so they can work together as one. What is undesirable is too much "wanting to receive" energy and not enough "I'll give" or "I'll share."

– SECTION TWO – EVIDENCES OF TWO OPPOSITE REALMS –

The Pawlowski Model Clarifies Religious Truths

I believe that by realigning scientific thought, many "godless theories" will disappear. This statement is not meant to demean scientists or question their extraordinary work, for truths pertaining to the physical world have sprung forth in great profusion during the past 150 years. The bulk of this information has blessed mankind immeasurably. But many have recognized that in the early days of the development of the quantum theory, some material was smoothed out in order to make the theory work. Instead of using sound scientific methods, wrong conclusions were formed by some who were unwilling to "wait out" or study questionable areas in greater detail. The patterns they saw were so seemingly beautiful that when the conclusions did not quite fit the facts, they used their own philosophical ideas to explain the discrepancies. Instead of acknowledging the blanks, they chose to work around them. It is conceivable that these inconsistencies originated from the small, but puzzling movement of electrons indicated in Bohr's chart. Since they could not be explained, an uncertainty factor entered the picture and exacting law was dismissed.

This lack of knowledge has greatly affected our lives, for by bringing probability-chance and uncertainty into the creation process, scientific philosophers have been led to speculate on the origins of life. Some have said there is no reality on the subatomic level and that an intelligent observer actually creates reality. I feel this puts a damper on our lives. And when told that evolution is what brought us to our present state, it is only natural to say within ourselves: "*Why do we need a creator if things happen by chance?*" and "*Why must we obey laws if there are no set laws?*" This type of thinking has had a profound effect in creating the "godless society" we now live in, for by rejecting exacting laws, they have rejected our Omnipotent Creator. By subtly downplaying the purpose of our existence and discounting the presence of an all-knowing God, science has eroded the very foundation upon which we must build.

The Spiritual Aspects of the Pawlowski Model of the Atom

The Pawlowski Model of the Atom is an incredible discovery and will be invaluable to the scientific community, but it has even more important spiritual aspects to focus upon. It was Mrs. Pawlowski's desire that this model be used as a testament to explain spiritual laws and concepts that have not previously been fully understood. Among these concepts are:

1. Creation is exacting, therefore must have a Creator.
2. For balance to occur, we need a spiritual as well as a physical body.
3. There is a spiritual as well as a physical death.

Have we been kept in the dark on such important concepts as these? Have there been answers, but we failed to recognize them as such? There have been answers, but not the kind we like to hear. From the beginning, prophets of God have been warning of the pitfalls and hazards of this physical world. They have given us spiritual laws so that we might know the consequences that come if physical laws are broken. But since most people do not like spiritual utterances, but rather solid physical proof, many have turned a deaf ear on that which cannot readily be seen. But where has this insistence for physical proof gotten them? Right back to square one! They know no more about their origins than they did in the beginning.

Could Science and Religion Be Separate Parts of the Same Electrical System?

Scientists have taken us into an era we never could have imagined. Technological marvels abound because of their great knowledge and skill. Yet we have no peace. It is unfortunate that with all their creative genius, they have not been equally effective in moving the world ahead morally and spiritually at this same rapid pace. Nor have the religions of the world been able to bring about a universal spiritual rebirth. I believe that if religion and science were to come together we would then have the full picture. Many will argue that this is impossible since religion and science are like oil and water that cannot be mixed. But let us consider this. Since the Pawlowski Atom reveals there must be two opposite energies for matter to be complete, could religion be embodied within the positive energy field that wants to give and share and science within the negative energy field that wants to receive? This would seem logical, for even though they are separate and distinct, when they come together; each becomes part of the whole system.

The Seen and the Unseen

Religion and science are like a lock and key. One is in the spiritual realm, the other the physical realm. Each is concerned with the individual, but since the spiritual motivates individuals into doing good works, it is the strong force that moves things forward. It is the part that "acts" while the physical is the part that is "acted upon." Both are necessary for the unit to work, but religion, the spiritual part, is the key that "acts" upon the lock. Scientists work with the physical elements of the earth, whereas prophets speak of an unseen world, close at hand, made of elements more refined and pure. It is the unseen spiritual part that unites these two worlds, because the physical world by itself cannot act on its own. Science currently calls this connection "string theory." Most people call it religion. Because spirit element cannot be seen with the natural eye, many dismiss its existence. But could spirit matter be what activates physical matter to give it life? Do spirit matter and physical matter need each other to be complete? I am confident the answers are yes.

TWO OPPOSITE FORCES — TWO OPPOSITE WAYS

We can assume there are two opposite forces outside our bodies since there are two opposite forces within our bodies. I believe it is fair to say that all have experienced the reality of these two opposites at work both from within and without our bodies (as well as throughout society as a whole), as sometimes a dark side appears, while at other times, a light. What could these forces be that are trying to affect our elements? Could one force be the Great Creator that religion has been telling us about, while the other force, Satan, His adversarial opposite? This would seem to be so, for one is building and enhancing people's lives (a plus +), while the other is tearing down and destroying them (a minus –). One is moving things forward by organizing matter (a plus +), while the other is doing all that he can to disorganize matter (a minus –). One is building and creating new worlds (a plus +), while the other does not have the ability to create (a minus –). These two forces cannot co-exist under such conditions, for the Pawlowski Atom reveals that for us to live in an electrical world that is seeking balance and equilibrium, positive and negative forces must work together for life to be sustained. This is where the problem lies, for mankind's challenge since the beginning has been that the negative force is always trying to overwhelm the positive force. To keep matter organized and working together as one, rather than in chaos, continues to be our test.

The Bible has much to say about these two forces, but for now, our focus will be mainly on the *works* of the Great Creator who moves things forward. Later, the works of the adversary, who pulls things back, will be discussed (see pages 89 to 98).

God is the God of Creation

The first sentence in the Bible declares: "*In the beginning God created the heaven and the earth.*" (Genesis 1:1)

This declaration boldly proclaims that God is responsible for the creation of this world and all things in the heavens above.

How did He accomplish such an awesome task? The Bible also declares that it was through Jesus Christ, His Only Begotten Son! "*God...who created all things by Jesus Christ.*" (Bible, Ephesians 3:9) It was he who organized the elements so they would work together as one! His role could be likened to that of construction foreman, overseeing and directing this magnificent building project, all the while, under the direction of God, who was his Father:

> "In the beginning was the Word, and the Word was with God...All things were made by him; and without him was not anything made that was made...And the Word was made flesh, and dwelt among us, (and we beheld his glory, the glory as of the

only begotten of the Father), full of grace and truth." (Bible, John 1:1, 3, 14)

And from Moses:

> "And worlds without number have I created; and I also created them for mine own purpose; and by the Son I created them, which is mine Only Begotten." (Pearl of Great Price, Moses 1:33)

Then from the Son of God himself, we read the following words:

> "Behold, I am Jesus Christ the Son of God. I created the heavens and the earth, and all things that in them are. I was with the Father from the beginning." (Book of Mormon, 3 Nephi: 9:15)

Did Jesus Christ create the world out of nothing (meaning ex Nihilo) or did he have materials to work with? The book of Abraham tells us:

> "And there stood one among them that was like unto God, and he said to those who were with him: We will go down, for *there is space* there, and *we will take of these materials, and we will make an earth,* whereon these may dwell." (Pearl of Great Price, Abraham 3:24, emphasis added)

Create means to cause to exist, to originate, to bring about and to organize. A chemist combines elements to bring about new compounds; a physicist is involved in the study of how matter and energy interact. To create something, both must have materials to work with, whether it be physical things or abstract things of the mind. There must be a starting point. Like the above scripture points out, "we will take of these materials, and we will make an earth." *Elements* are the starting point in the process of creation and Jesus Christ is the God that creates!

THE SCRIPTURES TESTIFY OF THE LAW OF OPPOSITES

The prophets have born testimony that Jesus Christ, under the direction of God, his Father, created the world. Is there any evidence in the scientific world that would lead one to believe this is true? There is, for the patterns discovered in the Pawlowski Atom give compelling proof that Jehovah, the premortal Christ, followed the *law of opposites* to accomplish such an awesome task. How do I know this? The Book of Mormon, a book of ancient scriptures, speaks of this opposition. It gives an account of a prophet named Lehi, (who lived 600 years before Christ) telling his son Jacob about the creation of things. He said: "...it must needs be, that there is an opposition in all things." He also said that God created both things "to act" and things to be "acted upon" and that "all things must needs be a compound in one, for should it be one body, it must needs remain as dead." This inspired scripture helps us understand the nature of God and the law of opposites upon which He operates. It also helps us understand the nature of science and the law of opposites upon which it operates:

> "For it must needs be, *that there is an opposition in all things*. If not so...righteousness could not be brought to pass, neither wickedness, neither holiness nor misery, neither good nor bad. Wherefore, *all things must needs be a compound in one*; wherefore, *if it should be one body it must needs remain as dead*, having no life neither death, nor corruption nor incorruption, happiness nor misery, neither sense nor insensibility. Wherefore, it must needs have been created for a thing of naught; wherefore there would have been no purpose in the end of its creation. Wherefore, this thing must needs destroy the wisdom of God and his eternal purposes...And now my sons, I speak unto you these things for your profit and learning; for there is a God, and he hath created all things, both the heavens and the earth, and all things that in them are, *both things to act and things to be acted upon*." (Book of Mormon, 2 Nephi 2:11-12,14, emphasis added)

What amazing evidence! The *law of opposites* is at work everywhere. When Lehi said there must be things "to act" and things "to be acted upon," he could not have described God's creations in plainer words than these, for to achieve His eternal purposes, He has organized matter so that a positive spirit entity and a negative physical entity can unite to become one entity. And science could not more clearly describe the relationship that exists between positive and negative elements; for with positive elements "acting" and negative elements being "acted upon," they, too, can become one entity. All matter is a unity of opposites and a compound in one. It is the action and reactions between positive and negative energy that is making the world go around.

With this new knowledge, the picture is now complete. Religion and science are separate parts of the same system, with religion coming from the positive, spiritual realm and science coming from the negative, physical realm. Science has been unable to give the correct answers to "whence came man" because it did not have the spiritual picture, yet the world has been slow to accept true religion because it did not have the physical proof. Both witnesses were needed for the picture to be complete. Which came first? The spiritual witness, since it is the stronger of the two. Also, God's hand could not have been revealed if the spiritual witness came last.

The Law of Opposites is the Governing Law

Matter has been organized so things which "act," the positive (+) can unite with things "to be acted upon," the negative (–). This combining of elements occurs when our spirit bodies enter our physical bodies. As long as the laws governing these two opposites are obeyed, they will complement one another: if not, they become opposites that work against one another. People then ask: "Is there a right and a wrong way or a good and a bad way, or does it just depend on which side you are on?"

The only way to answer this without being confrontational is to say that since our bodies are all made from the same elements, we are all under obligation to obey the same laws. We cannot dismiss this truth, for God's laws are immovable and unchanging. To understand the importance of law, the prophet Lehi said:

> "And if ye shall say there is no law, ye shall also say there is no sin. If ye shall say there is no sin, ye shall also say there is no righteousness. And if there be no righteousness there be no happiness. And if there be no righteousness nor happiness there be no punishment nor misery. And if these things are not there is no God. And if there is no God we are not, neither the earth; *for there could have been no creation of things, neither to act nor to be acted upon*; wherefore all things must have vanished away." (Book of Mormon, 2 Nephi 2:13, emphasis added)

Had there not been an opposition established in all things, there would have been no creation and nothing to "act upon." Matter would have remained unorganized and in chaos. And had the bonding power of the law not been established to hold all things together, there would have been no sin, no happiness or righteousness, for we would have been naught. There would have been nothing, had not God's grace organized the elements into a oneness of purpose.

How could Lehi have known that creation would be incomplete unless there were two parts: a part that "acts" and a part that is "acted upon?" Here was a man who had left Jerusalem over two thousand years ago to settle in the lands of America. Little was known of the minute, unseen particles of the physical world, yet he wrote words more attuned to an enlightened scientific age. They were revealed to him when man knew little of the physical world, so that at the end of time, they could testify that there is a God in heaven who is the source of all truth and that He rules and reigns over all things throughout the universe!

Lehi's words will stand as an unshakable witness of the divine mission of Jesus Christ. It was he, who under the direction of the Father created the world by organizing matter into a pattern of positive-negative opposites and it was he who brought power to the whole. Why did he do it? It was for the glory and benefit of man. Should we not give honor and show great reverence to him who hath provided such a gift?

TWO CREATIONS: HEAVEN AND EARTH - SPIRITUAL AND PHYSICAL

God refers to His creations as having two parts: heaven and earth or it could be said, the spiritual and the physical:

> "...Behold, I reveal unto you concerning this heaven and this earth; write the words which I speak. I am the Beginning and the End, the Almighty God; by mine Only Begotten I created these things; yea, in the beginning I created the heaven, and the earth upon which thou standest." (Pearl of Great Price, Moses: 2:1)

How did God create the earth? Since everything in the universe is made up of matter/element, when He created the earth, He took matter unorganized, and brought it into an organized state. He said, "We will go down, for there is space there, and we will take of these materials and we will make an earth whereon these may dwell..." (Pearl of Great Price, Abraham 3:24). We learn that this gathering together of the elements was during the preliminary stage: "And the earth, after it was formed, was empty and desolate, because they had not formed anything but the earth..." (Pearl of Great Price, Abraham 4:2). The earth was unable to produce because it was sterile and without life. To prepare the earth for man's use, opposites such as light and darkness, water and dry land, plant and animal, fish and fowl were established in order that life could be created and sustained.

On the first day God said: "Let there be light and there was light." (Pearl of Great Price, Moses 2:3). During the next five time periods, He said:

> "...Let the waters under the heaven be gathered together unto one place....Let the earth bring forth grass, the herb yielding seed, the fruit tree yielding fruit....Let there be lights in the firmament of the heaven, to divide the day from the night....the greater light to rule the day, and the lesser light to rule the night....Let the waters bring forth abundantly the moving creature that hath life, and fowl which may fly above the earth....And I, God made the beasts of the earth...and cattle...and everything which creepeth upon the earth..." (Pearl of Great Price, Moses 2: 9,11,14,16,20, 25.)

When the earth was ready to be inhabited, God said:

> "And I, God, created man in mine own image, in the image of mine Only Begotten created I him; male and female created I them...And I, God, blessed them, and said unto them: Be fruitful and multiply and replenish the earth, and subdue it and have dominion...over every living thing that moveth upon the earth." (Pearl of Great Price, Moses 2:27-28)

God then rested from His labors. It was the seventh day and the beginning of Sabbath day worship. "*Six days thou shalt work, but on the seventh day thou shalt rest*" (Bible, Exodus 34:21). The Sabbath is to be a holy day, set apart for worshipping God our Father and His Son Jesus Christ.

All Things Were First Created Spiritually

God created all things spiritually, before sending them to earth to enter mortal frames:

> "For in heaven created I them; and there was not yet flesh upon the earth, neither in the water, neither in the air...all things were before created; but spiritually were they created and made according to my word ...And every plant of the field before it was in the earth, and every herb of the field before it grew. *For I, the Lord God, created all things of which I have spoken, spiritually, before they were naturally upon the face of the earth.*" (Pearl of Great Price, Moses 3:5,7, emphasis added)

All Spirit is Matter

Spirit matter is the positive spark that activates negative physical matter to give it life. Even though spirit matter cannot be seen or discerned by the naked eye, nevertheless we have been assured by God that it exists: "*All spirit is matter, but it is more fine or pure.*" (Doctrine & Covenants 131:7) There is no such thing as ex nihilo or creation out of nothing, for all things are made up of matter. By following exacting laws, Jesus Christ organized existing matter into a dwelling place for man.

All Things Have Been Created in the Likeness of Its Own Kind

What we see in this physical world is the opposite of its heavenly counterpart. To know one, is to know the other. We are not reincarnated time after time into things not of our kind, for we will always be ourselves, separate and distinct from all others. It is the same with all living things. The physical entity will look like the spirit entity that enters to give it life. Plants will enter plants, animals will enter animals, fish will enter fish, and insects will enter insect's etc, for God said:

> "*Let the earth bring forth the living creature after his kind*; cattle, and creeping things, and beasts of the earth after their kind, and it was so." (Pearl of Great Price, Moses 2:24, emphasis added)

> "*That which is spiritual being in the likeness of that which is temporal*; and that which is temporal in the likeness of that which is spiritual; the spirit of man in the likeness of his person, as also the spirit of the beast, and every other creature which God has created." (Doctrine & Covenants 77:2, emphasis added)

All Things Fit Together Like a Lock and a Key

Heaven and earth are opposites—one gives, the other receives. The heavenly spiritual creation acts to give life, whereas the earthly creation is acted upon and receives life. When an earthly physical creation reaches a state of readiness, the spiritual creation can then unite with it. Otherwise it would remain "as dead." For creation to be whole and living, two parts must exist in opposing energy fields, to act and be acted upon. With one in a positive energy field and the other in a negative energy field, all things fit together like a lock and a key.

WHY DID GOD CREATE THE HEAVENS AND THE EARTH?

The book of Abraham, an ancient record written 4,000 years ago during the time of the Pharaohs, gives new insights into the eternal nature of God and His creations, as well as shed new light on our beginnings. In it's original form it came from the hand of the prophet Abraham. It is now part of a larger book called the Pearl of Great Price. In it, Abraham was told of the creation of worlds: of the laws, which govern the sun, the moon and the stars, and of the eternal nature of spirits. He was also told of man's pre-earth life and of the immensity of space being filled with intelligences ranging from the very small to the very great.

> "If two things exist, and there be one above the other, there shall be greater things above them...also, if there be two spirits, and one shall be more intelligent than the other, yet these two spirits, notwithstanding one is more intelligent than the other, have no beginning; they existed before, they shall have no end, they shall exist after, for they are...eternal. And the Lord said unto me. These two facts do exist, that there are two spirits, one being more intelligent than the other, there shall be another more intelligent than they: I am the Lord thy God, *I am more intelligent than they all*...I dwell in the midst of them all; I now, therefore, have come down unto thee to declare unto thee the works which my hands have made, wherein my wisdom excelleth them all, for *I rule in the heavens above and in the earth beneath, in all wisdom and prudence, over all the intelligences thine eyes have seen from the beginning*; I came down in the beginning in the midst of all the intelligences thou hast seen." (Pearl of Great Price, Abraham 3: 16, 18,19, 21, emphasis added)

These intelligences had not been created, nor made, but had always existed and were co-eternal with God. Like the elements in the immensity of space, they were part of the building blocks of creation. During a pre-mortal time, these intelligences were embodied within spirit element, but even then, each continued to be itself, separate and distinct from all others.

For What Purpose Were the Intelligences Organized?

The intelligences were organized to see if they would do all things whatsoever the Great Creator commanded them to do. He created worlds without number and laid out the laws by which all things were to be governed. Should they prove faithful in obeying His laws, they would be added upon and given the opportunity to further progress.

> "*And we will prove them herewith, to see if they will do all things whatsoever the Lord their God shall command them*; And they who keep their first estate shall be added upon; And they who keep not their first estate shall not have glory in the same kingdom with those who keep their *first estate*; and they who keep their *second estate* shall have glory added upon their heads for ever and ever." (Pearl of Great Price, Abraham 3:25,26, emphasis added)

The *first estate* of man was the time we lived as spirits in the presence of this Great Being who was "more intelligent than they all." Since it was He who begot the spirit bodies in which our intelligence dwells, He is our Heavenly Father as well as the God of Creation. He is the Elohim of old and the one Great God.

The *second estate* of man refers to this earthly existence. Even though we have remained the self-same entities, with each new step we have been "added upon." By taking on a physical body, our powers were expanded and new dimensions brought to our being. The importance of this second estate in which we now live cannot be overstated, for our place in the eternities depends upon what we do here. Keep in mind, (as stated in the previous scripture) that those who "kept not" their first estate were restricted and could not move forward to a greater glory. Here in mortality it will be the same. Those who "keep not" this second estate will likewise be restricted and unable to move forward to a greater glory.

A Plan Presented

Heavenly Father's intelligence was so extended and far-reaching that He had power over all things. Yet, because He was not jealous of His power and might, He ordained a *plan* whereby His spirit offspring could qualify themselves to become exalted, like unto Himself. Each would, in turn, leave His presence and come to earth to obtain a physical body and gain earthly experience. Each could progress toward perfection and ultimately realize his or her divine destiny of "eternal life." This plan, called the Plan of Salvation, was to assist His children while on earth. For this plan to be carried out, He needed someone to represent Him and supervise His work. He needed an intermediary and a messenger to connect heaven and earth. He needed someone to set a perfect example for his children to follow. He therefore called a Grand Council to present this plan, as well as choose this emissary who would represent him. There He asked:

> "...*Whom shall I send?* And one answered like unto the Son of Man: Here am I, send me. And another answered and said: Here am I send me. And the Lord said, *I will send the first.* And the second was angry, and kept not his first estate; and at that day, many followed after him." (Pearl of Great Price, Abraham 3:27-28, emphasis added)

Two personages responded to the Father's query: Jesus Christ, the Firstborn of our Heavenly Father's spirit children and Lucifer (Satan), also His son. Both gave a similar response. But there was a difference in the motives of these two sons:

> "And I, the Lord God, spake...saying: That Satan...is the same which was from the beginning, and he came before me, saying—Behold, here am I, send me, I will be thy Son, and *I will redeem all mankind, that one soul shall not be lost, and surely I will do it; wherefore give me thine honor. But, behold, my beloved Son, which was my Beloved and Chosen from the beginning, said unto me—Father, thy will be done, and the glory be thine forever. Wherefore, because that Satan rebelled against me and sought to destroy the agency of man*, which I, the Lord God, had given him, and also, that I should give unto him mine own power; by the power of mine Only Begotten, *I caused that he should be cast down*; And he became Satan, yea, even the devil, the

> father of all lies, to deceive and to blind men, and to lead them captive at his will, even as many as would not hearken unto my voice." (Pearl of Great Price, Moses 4:1–4, emphasis added)

The Savior's petition was one filled with love and obedience while Satan's was self-praising and designed to take away man's agency. We could not have progressed under Satan's force and coercion, for to become a self-governing entity, we had to be held accountable for our actions; not saved without effort.

Agency is at the Center of God's Plan

At the Grand Council, all of Heavenly Father's children were given the freedom to choose the Father's plan or to rebel with Satan. As with any choice, there is a consequence. Because Satan and his followers chose to rebel, they will never be privileged to have a physical body or to progress further:

> "...He (Satan) rebelled against me, saying, Give me thine honor, which is my power; *and also a third part of the hosts of heaven turned he away from me because of their agency*. And they were thrust down, and thus became the devil and his angels; And, behold, there is a place prepared for them from the beginning which place is hell." (Doctrine & Covenants, 29:36-38, emphasis added)

Since this premortal, first estate, there have been opposite choices and opposite ways. Jesus Christ was in complete accord with the Father's plan, therefore, was chosen to lead and direct His work. He would be God's spokesman and emissary, both in heaven and on earth. Satan rebelled against the Father's plan and from that time forth, sought to destroy the Father's work. This choice, made in the pre-mortal world, has made him an adversarial opposite to God. He proceeded to put out a counterfeit version of God's plan that continues to confuse Heavenly Father's children to this present day.

These two *spirit* children chose different paths and different ways. The Savior's way of giving, reaching out, and helping others is the perfect example of the right, positive element that wants to give and share (a plus +). Satan's way of selfishness and power grabbing is the perfect example of the left, negative element that wants to receive and be satisfied (a minus –). The Savior's path creates positive energy that builds the individual while Satan's creates negative energy that destroys the individual. With these two personages having such opposite natures, motives, and ways, the rest of Heavenly Father's children would be subjected to some very important choices here in mortality.

Many in the first estate were condemned and ceased to progress because they followed Satan's path. It will be the same in this second estate. We too, will be under condemnation should we refuse the light (truth) that was plainly manifest in the beginning:

> "Behold, here is the agency of man, and here is the condemnation of man; because that which was from the beginning is plainly manifest unto them, and they receive not the light. And every man whose spirit receiveth not the light is under condemnation." (Doctrine and Covenants, 93:31-32)

Our opportunities for further growth and development throughout the eternities will come as a direct result of our actions here on earth. In order to become masters of our own destiny, we have been given the freedom to act for ourselves. This is a most important aspect of Heavenly Father's work:

> "...Nevertheless, thou mayest choose for thyself, for it is given unto thee..." (Pearl of Great Price, Moses 3:17.)
>
> "All truth is independent in that sphere in which God has placed it, to act for itself, as all intelligence also; otherwise there is no existence." (D & C, 93:30)

To act for ourselves! That is the key to understanding why we are here. Free agency is necessary for us to determine our own course of action. By having our agency, we learn from our own experiences that obedience to laws that were established before the foundation of the world is the true source of happiness and joy. By choosing to follow that divine light, we choose that which was plainly made manifest from the beginning.

THE GARDEN OF EDEN: THE BEGINNING OF OPPOSITE CHOICES

Adam and Eve were the first of Heavenly Father's children to take up residence on this planet earth. He placed them in a beautiful garden, called Eden. There, amid the beauties of nature and the abundance of fruit of every kind, He gave them two commandments. They were to multiply and replenish the earth, and they could eat of the fruit of the garden *except* the fruit of the tree of knowledge of good and evil: for in the day they partook of that fruit, they would surely die. These two commandments were at odds with one another because they could not have children since they were not yet mortal and they could not become mortal unless they partook of the fruit of the tree of knowledge of good and evil. Adam and Eve could choose between two ways. They could remain in their innocent state or they could progress and become wise. One choice would stop their progression; the other would open up unlimited opportunities.

Because of the need for opposition, Satan was allowed to tempt mankind during this mortal probation. His first encounter was with Adam and Eve in the garden. There he persuaded Eve to partake of the forbidden fruit. This act of disobedience affected her very elements and opened her eyes to the reality that there was something other than peaceful bliss. When Adam saw what Eve had done, he too, partook of the forbidden fruit, thus enabling them to become as the Gods—knowing good from evil:

> "For God doth know that in the day ye eat thereof, your eyes shall be opened, and ye shall be as the Gods, knowing good and evil. And when the woman saw that the tree was good for food, and that it was pleasant to the eyes, and a tree to be desired to make one wise, she took of the fruit thereof and did eat, and gave also unto her husband...and he did eat. And the eyes of them both were opened." (Bible, Genesis 3:5–7)

Adam and Eve Should Be Held in High Esteem

Adam and Eve should be held in the highest esteem for the part they played in history's beginning scene. Had they not partaken of the forbidden fruit, they would have remained in the garden as innocent children, forever blocked from a higher, more blessed state of existence. In this state of innocence, they would have had no children; thus we too would not have progressed. They could not have become wise, for they knew no evil. Mortality would be a time of learning the consequence of opposite ways:

> "And now, behold, if Adam had not transgressed he would not have fallen, but he would have remained in the Garden of Eden. And all things which were created must have remained in the same state in which they were after they were created; and they must have remained forever, and had no end. And they would have had no children; wherefore they would have remained in a state of innocence, having no joy, for they knew no misery; doing no good, for they knew no sin. But behold, all things have been done in the wisdom of him who knoweth all things. *Adam fell that men might be; and men are, that they might have joy*." (B of M, 2 Nephi 2:22-25, emphasis added)

Rather than damn their progression, Adam and Eve chose to disobey a lower law in order to obey a higher law. By eating the forbidden fruit, they became mortal, thus giving them the ability to multiply and replenish the earth. Because of their fallen state, they had to leave their beautiful garden and enter the lone and dreary world; a world in which Satan had great power. Because a veil of forgetfulness had been drawn over their eyes, they could no longer remember their previous bliss. They had sown within themselves the seeds of sickness and death, therefore trials and sorrows became a part of their daily lives.

Heavenly Father did not want Adam and Eve to remain forever in their fallen state, therefore He provided a plan (the gospel) whereby they would be delivered from the undesirable consequences that had come upon them. It was the same plan as presented in the pre-existence. Since they could no longer remember their pre-earth state, it had to be taught to them again. Jesus Christ, the Only Begotten Son of the Father, was given this responsibility. By his voice and by the voice of all his prophets, he taught Adam and Eve, as well as all future inhabitants of the earth, the gospel from the beginning. This plan of redemption was embodied within a framework of law, with covenants and ordinances required of all that would embrace it.

> "Adam and Eve blessed the name of God, and they made all things known unto their sons and their daughters... *And thus the gospel began to be preached, from the beginning*, being declared by holy angels sent forth from the presence of God, and by his own voice, and by the gift of the Holy Ghost. *And thus all things were confirmed unto Adam, by an holy ordinance*, and the gospel preached, and a decree sent forth, that it should be in the world, until the end thereof; and thus it was. Amen." (Pearl of Great Price, Moses 5:12,58,59, emphasis added)

After the gospel was preached to Adam's posterity, they were given an opposite choice:

> "And Satan came among them saying: I am also a son of God; and he commanded them saying: Believe it not; and they believed it not, and they loved Satan more than God. And men began from that time forth to be carnal, sensual and devilish." (Pearl of Great Price, Moses 5:13.)

This pattern of deceit and opposition to the Father's work started in the pre-mortal life and continues to the present day. Even now, Satan says, "Believe it not," when Heavenly Father's plan is presented. But if we listen to his promptings we will be damned, which means our progress will stop. Just as a dam will stop the flow of water from going forward, should we proceed counter to Heavenly Father's will, our progression will also stop and we will be unable to move in a forward direction.

Adam and Eve were given a choice between living a higher law and a lower law. By obeying the higher law, they moved on to greater blessings. We, as their literal offspring, are being given the same opportunity to advance and become wise. Should we not therefore choose the better way?

MANKIND ENTERS THE WORLD TO BE TESTED

As the posterity of Adam and Eve, we too, are being given the opportunity of becoming wise and all knowing. Since we have also taken on physical bodies made of the same elements as the earth, we, too, are expected to obey the laws, which made these two bodies complete and whole. We must also be tested to see if we will do all things whatsoever our Heavenly Father commanded. Heavenly Father has been trying to inform us of these laws and what will happen if we do not obey them. On the other hand, Satan, the rebellious one, has been telling us there are no laws! Heavenly Father's message appeals to our spiritual, positive side, while Satan's to our physical, negative side. Heavenly Father's ways refine and ennoble us and make us want to be good and clean. They uplift and bring out our divine nature. Satan's ways debase, corrode, and appeal to our baser instincts and promote excessive gratification of physical needs. Should we listen to Satan's enticements, we would overburden our physical body with excessive negatives, thus disrupting our two bodies that want to work as one.

Could Our Conscious Physical Actions Upset the Balance With Our Spiritual Side?

When speaking of matter, we are speaking of energy, because matter is energy. Since our bodies are made up of two opposite energies, we are electrical beings that are continually converting matter into either positive or negative energy. Even the way we think is of great import, for positive thoughts create positive energy and negative thoughts create negative energy.

By extending this premise into every aspect of our lives, that which we see, hear, smell, feel, eat, or do will affect our bodies in either a positive or negative way. With our different body parts having the ability to process the positive and negative information we take in, we need to be careful that our actions are not detrimental. What if we were to allow a noxious substance to enter our mouth or nose that was harmful and poisonous to the spirit that is bringing us life? We would be allowing negative (–) matter/energy to enter. What if we were to allow our eyes to see obscene pictures that were offensive to our souls, or our ears to hear filthy jokes and degrading music? More negatives (–) would enter. And what if we were to allow our sacred reproductive parts to be carelessly handled and used for purposes other than what God intended? Even more negatives (–) and more trouble. With such an overabundance of negative energy, we would be in real trouble. Then we, as an inside spiritual force, would be allowing harmful outside physical forces to bring imbalance and harm to our bodies. The harmony that had existed between our spirit and physical body would be broken.

When Balance is Disrupted, Can it Be Restored?

At birth, the elements of our two bodies are much like an inert gas—balanced and whole. As we grow and mature this balance and equilibrium is disrupted by what we take into our physical bodies. When negative consequences occur, we begin to feel the affects of a broken law. We then want to do something to reverse our undesirable state. Can we ever return to a normal condition? Since the body is a living organism that has the ability to change, our elements can return to a state of balance and wholeness, but only by obeying the laws which govern them. These laws are encompassed within the restored Gospel of Jesus Christ, for embodied within its structure, is a plan that reveals God's perfect way. Its purpose is to bring healing to the sick and to deliver us from the ill effects of spiritual and physical death. This plan reveals the laws governing our energy driven world and outlines the parameters in which we must operate. The commandments it embodies teach us how to keep our spiritual and physical bodies working together as one. When the oneness is broken, we are taught how it can be restored. This plan makes known the course we must take in this physical world. If we submit to its conditions, it will bring peace to our souls, but if not, the balance and harmony we seek, will be illusive.

Why is man so out of balance? Spiritual things have been relegated to the back burner. Our nation's moral compass is fast eroding because many have lost faith that God exists. But He does exist, just as we exist! Even though our ability to honor Him publicly is being restricted, He is as real now as He ever was. To find Him, we must look to the spiritual realm, for here true religion will be found. True religion has the answers to "where we came from," "why we are here," and "where we are going." It has the answers that bring peace to the soul.

But how ironic, for even though our peace comes from the spiritual realm, it is within the physical realm that new evidence of God's reality is being revealed. Because of the Pawlowski Atom, the workings of the physical world can now be visualized, but more importantly, religion can now be understood in a new light. The law of opposites, upon which this atom operates, brings this new dimension to our thinking. We can now visually see how religion and science are separate, yet integral parts of the same system and why spiritual things must govern and oversee the whole. But to envision such a condition, one must put effort into learning how the atom is structured and behaves. This model of infinity helps us comprehend the never-ending nature of God's work and of the great potential that lies before us. By understanding its truths, we can envision a Supreme Being guiding the affairs of men who feels we are of great worth and worthy of His attention!

Yes, religion and science do go together and putting them side-by-side will yet reveal the full picture of why we are here. The Pawlowski Atom reveals that we have laws governing the elements of our bodies, just as satellites and spaceships have laws governing the elements from which they are made. These laws reveal the parameters in which our two bodies must operate. Once we know them, and the consequences that come should we go beyond them, we are without excuse. It is imperative we stay within the bounds God has set.

We Will Be Held Accountable For Our Actions

We are not just bystanders, idly watching others affect our lives; we are active participants in this great drama of life. We are in charge and should be correcting the things that need changing. If we are breaking God's laws, we must turn ourselves around and learn to obey them. In most life forms, instinct is the prime moving force, but man is different; his level of comprehension is unequaled among all God's creations. His ability to reason and think and understand the consequences of his behavior makes him "a little lower than the angels." Aristotle, the Greek philosopher, once said: "*Of all the wonders of the universe, the greatest is man.*"

As we look at this marvel of creation, we must remember from whence cometh our blessings. Do they come as a natural consequence of evolution or because we are the offspring of God? The answer is obvious: "*For in him we live, and move and have our being...We are the offspring of God.*" (Bible, Acts 17:29-2)

This is why God holds us to a higher standard and why we are so abundantly blessed. We are His offspring and He has created us in His image! But even though we have been created in a most marvelous way, we must realize there is an order within our elements that must be recognized and respected. The right, more refined spirit body that wants to give and share must govern the left, more coarse physical body that wants to receive and be satisfied, for God has said: "*It is the spirit that quickeneth, the flesh profiteth nothing; the words that I speak unto you, they are spirit.*" (Bible, John 6:63)

What is happening in the world today? Is life out of control because our spirit bodies are not controlling our physical bodies in their cravings for more physical things? Is a disproportionate amount of our time being centered on things that gratify our flesh? Are we losing control because our physical desires exceed our spiritual desires? Could these things be causing "Koyaanisqntsi," the Hopi Indian word meaning crazy life, life in turmoil, life out of balance? I think the answers are yes.

What is the Purpose of Our Creation?

The purpose of our physical creation is to have our spirit bodies added upon so that we can progress throughout the eternities! It is here, during the process of birth, that the spirit body enters the physical body so the two can become complete, as well as balanced and whole. Together they are capable of becoming perfected. In this way we can become like our Heavenly Father, for He, too, is made of spirit and physical elements, only more perfected and glorious than we can comprehend. He knew that if His children were to become like Him, they would have to leave His presence and become accountable for their actions. This is why free agency is such a vital part of His plan. We see within the elements, His grand design at work, as every person is given the freedom to choose that which will bring him life or that which will bring him death. This mortal existence is a time of testing, a time to perfect the union of positive and negative elements, and a time to learn that our spiritual side must become the head. Our Heavenly Father knew these two entities could become one, but not without effort. So we see a pull between the physical and the spiritual, the positive and the negative. Working together, they make beautiful music, but apart there are no eternal sounds.

COMMUNICATING WITH HEAVENLY FATHER

Our life here on earth is a time of spiritual testing and a time to prepare to move forward into a world of greater opportunities. God, who is the Father of our spirits has told us why we are here: "*That I may prove them, whether they will walk in my law.*" (Bible, Exodus 16:4.)

He has also said: "*There is no remembrance of former things.*" (Bible, Ecclesiastes 1:11) What would be the purpose in our having no recollection of "former things," and of having a veil of forgetfulness drawn over our eyes? It is so that we can learn to walk by faith during this short span in eternity:

> "*Dispute not because ye see not, for ye receive no witness until after the trial of your faith.*" (Book of Mormon, Ether 12:6)

What is faith? This next scripture tells:

> "Faith is not to have a perfect knowledge of things; therefore if ye have faith ye hope for things, which are not seen, which are true." (Book of Mormon, Alma 32:21)

There would be no need for faith if we were to remember the events of our pre-earth life. Faith must be developed in order for our spirits to become sensitive to the whisperings from on high. In this way, we can come to know and obey our Heavenly Father's will.

To have faith that Heavenly Father exists is the key to our success here on earth. Life changes once we know that He, who rules and reigns over all things throughout the immensity of space and who has all knowledge and control over the elements, is aware of us and wants us to succeed. He has said of His work:

> "For behold, this is my work and my glory – to bring to pass the *immortality* and *eternal life of* man." (Pearl of Great Price, Moses 1:39, emphasis added)

Immortality means being enabled to live forever with spirit and physical bodies inseparably connected and *eternal life* means to live the same type of existence as Heavenly Father lives. His work is therefore an *ongoing* process that will prepare us for a higher state of existence than that which we now enjoy.

Methods of Communicating With Heavenly Father

We must learn to recognize Heavenly Father's voice here in mortality so that we can communicate with Him and be taught in His ways; otherwise, how will we know right from wrong or which path to take if we are not given directions? It is through the following six ways that He instructs and guides us along the path of life:

Light of Christ: The light of Christ is given to each individual when entering this world so that all might distinguish right from wrong.

Holy Ghost: The Holy Ghost is a member of the Godhead. He remains a personage of spirit so that he may communicate with our spirits. He opens our minds to our Heavenly Father's will through whisperings, promptings, and impressions.

Gift of the Holy Ghost: This gift is given to a person after he or she has been baptized a member of the Church. It is bestowed by the laying on of hands by one who has the authority to bestow it. It teaches and instructs the individual in the ways of the Lord by giving personal and daily direction. It is a spirit to spirit communication that stays with a person as long as he or she remains worthy.

Prophets: These are men who have been endowed from on high to speak for Heavenly Father and His Only Begotten Son. They are their mouthpieces. They give direction on what we should or should not be doing in our lives.

Scriptures: These are the recorded words of the prophets and have been passed down for 6,000 years. They contain the laws and commandments of God.

Prayer: Personal prayer is how we communicate with our Heavenly Father. We ask for His direction and assistance and thank Him for what He has done, but we do it in the name of Jesus Christ, who is the mediator between God and man.

Each of the above connects the physical realm to the spiritual realm, thus making it possible to have a spirit to spirit communication with our Heavenly Father.

This contact is important, for just as parents here on earth are apprehensive when sending their children out into an uncaring world, Heavenly Father also felt great concern when sending His children down to earth. He knew that because of Satan's rebellion, many would be led astray. Just as earthly parents want to be assured of their children's safety, He too wants to communicate with us and help us while we are here! Do you think we could have sat in a heavenly library reading a book on mortality and truly understood? By having daily prayer and scripture study we learn how to be guided by the whisperings of the spirit. We learn how to resist Satan's temptations and move forward into the eternities! This spirit-to-spirit communication with our Heavenly Father is how we progress and become self-governing and capable of making choices that will bless not hinder our life. It is how we form a partnership with Him. He says, "Pray always, and I will pour out my spirit upon you." (Doctrine & Covenants 19:38)

THE UNIVERSE IS GOVERNED BY LAW

Law sets the parameters in which we must operate and restricts our actions so that order is maintained. When we think about law, our minds generally focus on four different types:

1. God's law
2. Man's law
3. Natural law
4. Moral law

God's law is a commandment or revelation coming from God—a divinely appointed order or system. *Man's law* is a rule of action or conduct laid down or prescribed by someone in authority—an edict or decree of a ruler or government as well as the whole body of rules regulating and controlling the individuals of a state or community. *Natural law* is a formal statement of facts invariably observed in natural phenomena. *Moral law* pertains to or is concerned with right conduct or the distinction between right and wrong. (*American College Dictionary*)

Law brings order to the individual and society by eliminating confusion and chaos, but all good laws rest on the foundation of God's laws. Though some people try to dismiss this reality, they cannot, because natural law, moral law, and scientific law are nothing more than manifestations of His laws at work. Some critics may look upon God's laws as being of little value and of no consequence, as something that only the ill informed continue to follow. But these critics should not be so content in their assessment, for no one can escape God's laws or the consequences they generate, for His laws are embodied within all the elements throughout the universe!

The Law Originates Within the Structure and Behavior of the Smallest Things

At the beginning of this book Rick Gore was quoted as saying, "In the structure and behavior of the very smallest things, lies the explanation of all matter." He was right, for the Pawlowski Atom has shown that once a pattern has been established in the smallest things, as they develop and become larger and more complex, their structure and behavior will naturally follow the same pattern that was set up in the beginning. This pattern could be called the law, for it governs and sets the boundaries in which all things must operate. Because it is embodied within the elements, it is immovable and unchanging.

While learning about the Pawlowski Atom, the following seven characteristics kept standing out in my mind as being important in determining the law. I'm sure there were others as important, but, to me, because they were always surfacing when the chemistry and physics parts were being explained, I felt they *were* the set boundaries in which we must operate.

1. *We are being governed by an exacting "law of opposites."* This was obvious, for how else could the 92 natural elements be found in two separate energy fields and working together as one, unless there was an exacting law that caused this division? The three key factors in organizing matter into positive and negative fields of energy are a polarity barrier, a figure eight electron path, and six "sliders." These invisible things cannot be seen with the natural eye, but their influence affects everything—even the way we measure time and seasons.
2. *The right side is the governing part.* As the atom expands outward into ever widening rays, the unusual movements in the alkali metals and alkali earths determine where the next electrons will be seated. The increased activity coming from this "slider area" causes the right side of the atom to be dominant. This area could be likened to its main terminal or master brain. There is no place like it in activity and energy levels. It is the governing part that moves things forward. Bohr's electron distribution chart confirms the existence of this unusual activity.
3. *Positive and negative energies are of equal importance, but positive is the stronger of the two.* This is not as contradictory as it may seem, for how else can power, magnetism, and light be generated unless the two fields are working together as one, yet how can the two move forward if both are of equal strength? The irregular movements and extra "sliders electrons" in the upper right quadrant creates more energy, therefore the right field moves things forward. If the two fields were equal in strength, there would be no movement. By being the stronger of the two, the positive can then act upon the negative. We see this same strong force-weak force pattern throughout all creation.
4. *All life forces are moving from right to left.* The first electron in the building of the atoms/elements is seated to the right of the polarity barrier, the second to the left. This gives evidence that energy is moving from day to night, east to west, positive to negative, because all the atoms/elements are following this same basic right to left pattern of opposites. This movement of energy would then be reflected throughout the universe.
5. *Balance and equilibrium are inherent within matter.* Had there not been a 2-8-8 pattern organizing the electrons along the eight rays of the atom, first to the right of the polarity barrier, then to the left: the balance between two energy fields could not have been established. And had the electron path not gone from an outer shell that expands to an inner shell that contracts, equilibrium could not have been maintained. It takes both for matter to be balanced and whole. Because of this opposite structure, we are therefore under obligation to establish this same balance and equilibrium between our spiritual and physical bodies.

6. *We must "give and receive" as well as "act and be acted upon."* These characteristics must become part of our lives because the elements of our body have been organized following this pattern. The positive spirit body "acts." It is the giver of electrons. The negative physical body "is acted upon." It is the receiver of electrons. By working together, the two remain balanced and whole.
7. *Freedom (expansion) yet restriction (contraction) is built into the elements.* Why do we have freedom, yet restriction in our lives? I believe it is a reflection of what is happening within the atom, for when electrons enter the right, positive energy field, nothing restricts their energy, for they are alone, and have nothing to "act upon." But when the electron path goes over the polarity barrier and electrons/elements start forming in the negative energy field, a restriction is set up because the magnetism between positive and negative is now in place. This magnetic attraction causes the negative field to be drawn in and restricted and limits the boundaries in which it can operate. It must then be "acted upon." And with the electron path moving from a higher shell that expands to a lower shell that constricts, wouldn't this expansion, yet contraction be reflected in the world around us? And wouldn't the alternating between positive and negative energy fields be causing our hearts to beat out, then in, and the tides to go out, then in? If all the atoms are beat to the same drumbeat, this movement of energy could very well be the rhythm and heartbeat of the universe.

How do These Laws Within the Elements Affect Us Personally?

We consist of two bodies, one spiritual and one physical; one positive and one negative. We learn from the atom that these two bodies are designed to work together as one. But we also learn that the spiritual body must be the governing part that oversees the whole. It must be the stronger part, otherwise, the physical body that is always wanting to be filled and satisfied, will overwhelm the spiritual body with negative energy. We also learn that the laws which govern our bodies are real and exacting, therefore, we cannot dismiss them or act as if they do not exist—and that freedom of choice is found in the positive spiritual realm, while restrictive laws in the negative physical realm. We learn that matter and energy are interchangeable, therefore, as electrical beings, we are continually converting matter into positive or negative energy. This is why it is important that we produce more positive energy than negative energy, for everything we see, hear, smell, feel, eat, or do, affects our bodies in either a positive or negative way.

We are Being Governed By a Law That Creates Cause and Effect

The *law of opposites* is the law that holds things together. Because it embodies the *law of balance and equilibrium*, it is also the law that creates cause and affect. Negative electrons bond to a positive nucleus, thus holding the atoms together. Positive atoms bond to negative atoms, thus creating new compounds. Magnetism is created when positive and negative matter unites. By bringing "positive and negative forces together," all things become a "compound in one." Men cleave to women, thus creating life. Plants grow when balance between positive and negative energy is maintained. Life continues as long as a positive spirit body cleaves to a negative physical body. But this oneness can be maintained only when the two work together as one. When the oneness is broken, life's power source stands in jeopardy and harmful consequences emerge. Divorces occur, people get sick, plants die, machines break down, and power goes off. It is a matter of life and death, for the elements hold the law.

When God's Laws are Broken, There are Penalties to Be Paid

We may not know or even understand the ramifications of some of man's laws, but when broken, we soon learn that ignorance is no excuse and we most likely will be punished for our error. But what about God's eternal laws? If they are broken, will our innocence shield us from the consequence of broken law? No, for they naturally occur within our elements when they are being broken!

But there is a difference in innocently breaking the law and deliberately breaking the law. We might innocently transgress the law and bring its undesirable consequences upon us, but we are not sinning in the eyes of God, unless we violate that which we know to be true. Sin occurs when we violate our own conscience—when we purposely and deliberately do the opposite of what we know to be right.

Society has become so accepting of sin that many do not think they are sinning when they knowingly break God's laws. They may suffer their undesirable consequences, yet they do not equate their problems with sin. But a person's acceptance of sin does not change the fact that sin is breaking the law. Sin brings undesirable consequences that diminish everyone because sin is a negative (–) that takes away.

Sin is such a secret thing that many times it is kept hidden from the world. But this desire to hide, so that no one will ever know, is an indication that man does know when he is sinning. Unless caught, he thinks, "All is well, life is great, and no one will ever know." But time has a way of changing this picture of contentment. As surely as night follows day, undesirable consequences will emerge because sin upsets the balance between two energies that should be working as one.

Perhaps the following analogy will help us better understand what happens when we sin. Suppose we have purchased a machine. The instructions clearly state how the machine should be run. We are told in bold letters *not* to turn the machine on high speed until the motor has had time to warm up. Instead we choose to disregard the instructions and immediately turn the machine on high speed. Now the machine will still work, but internally things are beginning to prematurely break down. We cannot see the change, we cannot feel the change, but after a time of sustained abuse, the machine will no longer work. Our physical body operates in much the same way. Our Creator has told us what we should and should not be doing to keep it spiritually strong, but when we knowingly or unknowingly disobey His instruction, it too will suffer the natural consequence of broken law.

With the growing acceptance of sin, we see just such a breakdown in society prevailing. This should not be too surprising, for when God's laws are tossed aside, man's laws stand in jeopardy because there is no moral base upon which to establish law. Anarchy will reign when man feels he is a law unto himself and has to answer to no one. Law enforcement agencies are then taxed to the limit trying to enforce the law and protect the people. It is evident that with such a large population, it is imperative we return to obeying God's laws, for how else can our efforts succeed?

Rejecting God and His Laws Creates an Irreligious State

Since the spirit and physical world are inextricably bound together by a law of opposites, should one part be missing or weakened there will be problems. Unless the two work together as one, maximum power, magnetism, and light cannot be produced. This imbalance between the spiritual and physical is the pressing problem of our day, for as the moral structure of our communities crumble at an ever-accelerating pace, we hear dissident voices telling us that the problem could be solved with more police, more public welfare, and better schools, more jobs, and less racism. Little is ever said about needing more spirituality, more righteousness, or more God-fearing citizens.

This lack of understanding the importance of the spiritual is understandable, since over the years the efforts of many have been ongoing and unrelenting in trying to break our ties with the spiritual realm. They have justified their position by claiming the Constitution says there must be a separation of church and state. This has been misconstrued, for religion was not even mentioned in the Constitution. It was after the Constitution was ratified that amendments were written as the need arose. Article 1 of the Amendments was then added as a protection for the people against a state-sponsored religion.

The founding fathers did not intend to stifle religion. They were God-fearing men who acknowledged the existence of a Supreme Being and who recognized His hand in establishing the new republic. Their families had come from countries where the state had embraced a specific religion—even using public monies to sustain that religion. The men who wrote the Constitution wanted to make sure all religions were given the freedom to exist and that each individual could worship God according to his or her own conscience.

This twisting of the Constitution has considerably weakened the spiritual values in our communities, for religion is important. No matter which religion we choose to embrace, its teachings should be the moral foundation and restraining influence in our lives. If children are not taught to pray and show reverence to Him who is the giver of life, they see no need to submit to any man. This disrespect for a supreme authority has now permeated the thinking of vast numbers of people and has caused the spiritual values of our society to be dramatically downplayed and dismissed. But how could it be otherwise, for when the presence of God and His laws are rejected, the people then readily accept sin.

How can this godless thinking be reversed? By combining the physical and spiritual, as was done in the past. Although some speak in strong unbending terms on the separation of church and state, they are, in reality, compatible, complimentary opposites that are different, yet, necessary for balance to be brought to the whole. With nation contending against nation, governments on the verge of bankruptcy, and corruption in high places, the need for such a union is obvious. And with irresponsible people fostering their wares upon a public that is tired of sleaze, there is no recourse for such abuses because the perpetrators have turned off their spiritual barometers.

To correct this imbalance, governments, too, must form a partnership with God. They, too, must have a spiritual side that restrains their physical side from becoming excessive. This does not mean the endorsement of a specific religion, but rather in bringing together individuals of proven integrity from all religious persuasions into an entity that governs and oversees the physical entity. With such a foundation upon which to build, its citizens can then publicly acknowledge their divine Creator without fear of criticism and censure and can go forward with full confidence in asking for His direction. It is evident from the Pawlowski Atom, that without a strong spiritual side, governments cannot expect to prosper in this physical world, for physical entities alone have no life! Like the pattern of positive and negative opposites within the atom, *the spiritual and physical must be separate, yet working together as one*, otherwise, power, magnetism, and light cannot be brought to the whole.

Diversity Within the Law Gives Freedom, Yet Restriction

What if all men were to repent and live God's laws? Would this make every person the same and the world an *un*interesting place in which to live? I think not, for people would still have their individual likes and preferences. Those who like to ride horses would still like to ride horses, and those who like to play baseball would still like to play baseball, and those who like to travel would still like to travel. Since freedom with restriction is at work within our elements, this same freedom with restriction would manifest itself in our lives. Should everyone repent of their misdeeds and obey God's laws, it means these same people would live righteous, moral lives and deal justly with their neighbors. They would treat others with kindness and be true to their spouses. They would be honest in all things, and integrity would be their most prized possession. Last, but not least, they would recognize that God is the source of their blessings and that offending Him would be a sin.

WHY DO WE ASSOCIATE SATAN WITH THE NEGATIVE?

We daily experience the desirable consequences of a world that has an east-west, day-night, hot-cold, right-left, soft-hard, and sweet-sour. In fact we cannot imagine life without them. These opposites are composed of two parts that complement one another. They work in harmony to bless our lives. But other opposites are not so desirable; things such as life-death, forgive-avenge, virtue-vice and love-hate. In contrast to desirable opposites, these are in opposition to one another. The energy from desirable opposites work and pull things together, ----->...<----- whereas the energy from undesirable opposites work against one another by pulling things apart, <-----...----->. One has balance and harmony; the other is out of round. One creates contentment, the other discord.

Desirable opposites come from obedience to natural law. Undesirable opposites come from disruption of natural law. As an example of the differences of these two opposites, light and darkness describe the natural workings of the physical world. But light and darkness also describe the consequences that come from good and evil. Positive and negative describe the two forces that are holding the world together. But they also describe good and bad behavior. What makes the difference? *The normal state* of the elements *is one of balance and equilibrium.* Anything that hinders, impedes or thwarts this oneness and does not bless and enrich our lives is undesirable.

Satan is the Disruptor of Natural Law

Who would want to disrupt this balance and equilibrium and defeat God's purposes? It is Satan, the rebellious one, and the spoiler from the beginning. By provoking unstable conditions among the peoples of the earth, he is momentarily hindering God's work and winning a few battles. But he has already *lost* the war, for we learn from the atom that positive power is greater than negative power. It is what moves things forward. It is the governing factor that oversees the whole.

Why do we associate Satan with the negative? It is because *he works through our physical senses rather than our spiritual senses.* He is not the physical, for the physical is good, but he hinders God's work by tempting us to go beyond the bounds that God has set for this physical world. Satan is sometimes called chaos, for when he succeeds in enticing us, our elements become disorganized from taking in too much negative energy. The balance between our spiritual and physical energies is then broken.

We look on negative as bad because it takes away (–) from our lives instead of builds (+). But negative per se is not bad, for if there were only "things to act" (the spirit body) and not things "to be acted upon" (the physical body), God's creations would be incomplete. The key is to keep them balanced and working together as one.

Unnatural State vs. Natural State

When man disrupts God's law, he goes against the natural state and supports Satan's work. This is why we are having problems in the world. God has created it one way, but so often man insists it be another way. But man cannot have it his way because problems emerge when God's natural laws are being violated. Under Satan's rule, our lives would be in a state of chaos.

Following are a few examples of how Satan is trying to disrupt the natural state:

Unnatural State–Same Gender Union — ***Natural State***–Opposite Gender Union

Males were created to be fathers and females to be mothers. This is the natural state; therefore, gender is an essential part of God's plan. The individual characteristics we bring from our premortal life, is part of His eternal purposes. Men and women have a natural built in attraction for one another, with the male giving the seed of life and the female receiving it. The Pawlowski Atom reveals that this attraction is inherent within the body, for like a magnet, opposites attract. One body (the positive) "wants to act" while the other body (the negative) "wants to be acted upon." This sets the pattern for all God's creations and, to do otherwise, is unacceptable to Him because it runs counter to natural law.

God is the God of Creation and putting positive and negative energies side by side, yet opposite one another, is how He creates. To want something other than this would be an odd, peculiar, and strange aberration. Should two male "wanting to act" (+) entities try coming together, they would naturally repel—the same with two female "wanting to receive" (–) entities. It is the lust of the flesh that brings such things together, not natural law. When the sacred act of procreation is abused and degraded, there will be negative consequences because the law within the elements is being broken.

Unnatural State–False and Dark — ***Natural State***–Truth and Light

Truth is reality in its purest form. Falsehoods distort reality. This is why lying goes against the natural state. It is founded upon deception and causes confusion in people's lives because they deny things they know to be true. A drug user will deny he is using drugs, an adulterer will deny he is being unfaithful, a liar will deny he is lying, and a thief will deny he has stolen. In this unnatural state, life becomes confusing for the one lying, as well as the one being lied to. What could be more confusing than to make a decision, only to find the information upon which you based your decision was a lie?

Those who think lying is a small thing should reconsider, for this "little" sin is the foundation upon which all others build. This is why Satan is called "the father of all lies." His work is the ultimate deception. He knows that sin can be kept secret as long as it is sustained and covered by a lie. The Lord says: "*Wo unto the liar, for he shall be thrust down into hell.*" (Book of Mormon, 2 Nephi 9:34). I remember my surprise, when first reading this scripture, for at the time it seemed such a harsh judgment for something so small. I have since come to realize why it is so severe. Consider the heartache lying has caused since the beginning and think of all those who might have accepted the truth, had they not been deceived. Life has taught me that lying is a big sin, not a little sin, for it goes against the natural state that God has established.

Unnatural State–War and Desolation — ***Natural State***–Peace and Prosperity

History has shown that war and desolation are not the natural order of things because it disrupts people's lives. Its devastating effects create unhappiness and sorrow as well as grief and pain. Even though most men throughout history have wanted to live in peace, there have been many, who like Satan, have desired power and dominion over others. But this is not the natural state. The unrest they initiate eventually leads to death, chaos, and ruin (–). Untold millions have felt the stinging blows of these despots of hate, therefore, at the time of judgment, those who have caused this condition of devastation and ruin will be dealt with in a very harsh way.

Unnatural State–Pollution and Decay — ***Natural State***–Clean and Pure

We all suffer in varying degrees from the pollution of the physical world. Our air, water, and land are in many instances, being poisoned by man's disobedience to natural law. We spray our crops with harmful pesticides and add chemicals to foods, which should remain pure. This contamination is affecting the very elements of our physical body, causing discomfort and even death.

But far worse than the pollution of the physical body, is the pollution of the spirit body. Its consequences are eternal and more devastating than physical death ever could be. We see this spiritual pollution attacking the very heart and soul of America. Visual images in the media have become increasingly bold in portraying a new way of life that promotes sin. The devastating effect this is having on our lives seems to be never-ending. How can the young get their bearings in life when such things are allowed to be shown? A society that has become too liberal in their thinking has created this chaos. But what can we expect? We cannot put garbage into our minds and expect something beautiful to emerge, for that would be contrary to natural law. We are now seeing the fruits of this mind altering bombardment of sin.

Unnatural State–Worldly lifestyle — ***Natural State***–Clean lifestyle

When our actions go beyond the bounds that God has set, we eventually have problems. When we smoke and drink or eat the wrong foods, we have problems. When we commit adultery or lie, cheat, and steal, we have problems. When we take drugs, or look at the wrong type pictures and listen to the wrong type music, we have problems. This is because the excessive negative energy they generate puts too heavy a burden upon us. Since we are taking in too many negative atoms that lack electrons, problems occur. We know this is true because the whole world is being overwhelmed with their negative consequences. Think how much our world would gain if positive energy were instead created and the laws governing our bodies were not being broken.

Is the Unnatural State Winning The Day?

Never before have there been so many choices or have ideas been communicated in such a universal way. Television and movies, magazines and newspapers, and videos and the Internet bombard us with a carnal and sensual way of life. Lifestyles are promoted that are degrading, coarse, and excessive. Yet society seems to no longer view them as insulting or forbidden. We should each ask ourselves, "How do I feel about these new trends? Am I embracing or rejecting this negative, pleasure seeking, ever wanting more, type of lifestyle?" This is an important test, for surely as the night follows the day, participation in the excesses of the world, will draw power from the spirit body that is bringing us life.

When a society is in the process of decay, there tends to be worship of the false gods of wealth, idleness, luxury, and pride. Little thought is given to how these excesses affect men's spiritual natures. Plato, the ancient philosopher, wrote of this condition:

> When the *divine portion* began to fade away, and become diluted too often and too much with the *mortal mixture*, and the *human nature* gained the upper hand, they then, being unable to bear their fortune, behaved in an unseemly manner, and to him who had an eye to see, grew visibly debased, for they were losing the fairest of their precious gifts; but to those who had no eye to see the true happiness, they appeared glorious and blessed at the very time when they were full of avarice and unrighteous power. ("The World's Last Mysteries," pg. 14, *Readers Digest*, emphasis added)

Are we today in danger of losing "the fairest of our most precious gifts," our spirituality? We are the greatest of all God's creations, so why are we lowering ourselves to such an irreligious state? We are His children and have been given the ability to reason and think on a higher level. Do we not know that holiness is an attribute of God and that no unclean thing will be received in His kingdom?

> "And he (God) doth not dwell in unholy temples; neither can filthiness or anything which is unclean be received in the kingdom of God; therefore I say unto you the time shall come, yea, and it shall be at the last day, that he who is filthy shall remain in his filthiness." (Book of Mormon, Alma 7:21)

It is Here That We Learn to Live God's Laws

This earthly existence is the time to learn to live God's laws and prepare for eternity. It is here that we learn to be spiritually, not physically oriented—to master our weaknesses and not be a servant to enslaving ways. Life was not meant to be an inward and ever taking sort of adventure, but rather a positive one of giving and sharing. For too long now, many have carelessly gone about doing whatever suits their fancy, never seeming to care how it affects other people's lives. They take no thought of how their actions influence others, just as long as it pleases them. This negative (–), ever wanting to be satisfied philosophy, comes from Satan. The key to correcting the imbalance brought on by sin is to obey God's laws and to serve others. This truth cannot be dismissed.

Unlimited Freedom On the One Hand—Restriction On the Other Hand

Does the Lord expect us to go forward like little tin soldiers that are restricted at every turn? It is a paradox, since we do have bounds in which we must operate, but we also have unlimited freedom to go forth on our own initiative. This dual nature of possibilities can be best illustrated by looking within the atom. The structure of the atom is such that when the positive energy field is being formed, its energy is unrestricted because, as yet, there is no opposition. However, an immediate restriction is set up once the negative energy field is being formed because the magnetism between the two fields is now in place. It is as if the negative field grabs hold of the positive field and puts on the brakes.

This braking of energy could very well be causing negative matter to be coarse and grainy and positive matter to be smooth and light. Although the two fields are dissimilar and seem incompatible, they are opposites that complement one another. Could it be the same with our two bodies? Could one body be reflecting the characteristics of the positive energy field, while the other, the negative energy field? Could one body be smooth and light, while the other coarse and heavy? And could religion, free will, and creativity be found within the right positive field and restriction within the left negative field? This would seem self-evident, for our spirits have the freedom to soar to the highest levels and to dream dreams that stretch into infinity. Here we can feel the presence of our Heavenly Father and catch a glimpse of a glorious world that we cannot yet see. But then we must come back to the realities of the earth, and like the negative field that is being restricted, we must then operate within the bounds God has set.

Our spirit bodies have the freedom "to act" whereas our physical bodies are restricted and have to be "acted upon." We are therefore given the freedom to create our own consequences. We can choose either good or evil. Good choices take us forward into infinity (+); bad choices, backwards into a diminished state (–).

Sin Becomes So Common Place

Why do we accept sin so readily? Alexander Pope taught a great truth:

> *Vice is a monster of such frightful mien, As to be hated needs but to be seen;*
> *Yet seen too oft, familiar with her face, We first endure, then pity, then embrace.*

The deterioration of our value system has shown this to be true. Our will is weakened when sin becomes commonplace. But make no mistake, sin is still sin no matter how familiar its face. It is evident, when looking at society's ills, that we are going to have to change our attitude toward sin or man's unbridled ways will eventually destroy us. Who can argue that too many minuses (–) within a bank account creates havoc. It is the same with us. A path of self-indulgence creates an overabundance of negative energy. By brushing aside God's laws in favor of Satan's ways, we are taking a path that is unacceptable to God. To stop the flow of excessive negative energy going into our systems, we must return to the natural state that produces and builds rather than the unnatural state that consumes and destroys.

WILL WE RESIST SATAN'S DAMNING WAYS?

Many do not acknowledge that Satan is real or that he motivates people into doing evil things. They do not understand that he is working in direct opposition to the Father and the Son. This is understandable, for his methods are so cunning that he can carefully lead us down to hell with our hardly ever knowing it:

> "For behold, at that day shall he (Satan) rage in the hearts of the children of men, *and stir them up to anger against that which is good.* And others will he pacify, and lull them away into carnal (worldly) security...and thus the devil cheateth their souls, and leadeth them away carefully down to hell. And behold, others he flattereth away and telleth them there is no hell; and he saith unto them: *I am no devil, for there is none*—and thus he whispereth in their ears, until he grasps them with his awful chains, from whence there is no deliverance." (Book of Mormon, 2 Nephi 28:20-22, emphasis added)

Oh, the awfulness of such a prospect! Surely we must come to recognize Satan and his ways. Because his work takes away instead of builds, it leaves all that it touches in a lessened, weakened condition. He persuades us to destroy the law, never mentioning that in the end our disobedience will destroy us. His work brings darkness since light and truth are taken from individuals and nations that persist in embracing his ways. When Satan persuades us to commit sin, it is deceit, for should he operate in the light it would reveal his true self.

Enticed By One or the Other

It is obvious from the opposites we see at work in the world today, that we are being enticed by two separate beings with opposite goals and opposite ways. Since the word entice means to draw on by exciting hope or desire and to lure, attract, and tempt with decoys, it should not be too hard to distinguish between the two. When Heavenly Father entices us, he tries to draw us to Him and His ways, with eternal life as the reward for our obedience. When Satan entices us, he tries to lure us with decoys that tempt us into doing things that prey on the weaknesses of the flesh and that go beyond the parameters that God has set. Depending upon which spirit we choose to heed, we receive either freedom or bondage, for strength of character, peaceful feelings, and uplifting thoughts come from Heavenly Father, whereas, weakness of character, agitated feelings, and degrading thoughts come from Satan.

One thing we must never forget in our making of choices, Satan can do nothing but tempt us. He can entice us, but he cannot make us do his bidding:

> *"Wherefore, the Lord God gave unto man that he should act for himself.* Wherefore, man could not act for himself *save it should be that he was enticed*

> *by...the one or the other*....they have become free forever, knowing good from evil; to act for themselves and not to be acted upon." (Book of Mormon, 2 Nephi 2:16,26, emphasis added)

We are free to choose between Heavenly Father's way or Satan's way. We can choose to be spiritual or carnal, clean or filthy, humble or arrogant, compassionate or indifferent. We can choose to love rather than hate, forgive rather than avenge and build rather than destroy. We can choose virtue over vice, freedom over bondage or law over anarchy, for:

> "Men are free according to the flesh;...And they are free to choose liberty and eternal life, through the great Mediator of all men, or to choose captivity and death, according to the captivity and power of the devil; for he seeketh that all men might be miserable like unto himself." (Book of Mormon, 2 Nephi 2:27)

Satan has many ways to lead us astray. He does it through the persuasion of others and by subtle seduction, by glorifying accepted sins and upholding false traditions; by extreme peer pressure, and by withholding pertinent facts (lying). Each of these persuasions come from an outside source. But should we be letting others determine how we will act? Of course not, for we must learn to act for ourselves and not be acted upon by others, otherwise, we cannot build up a resistance to evil.

This resistance to evil is mandated by our very elements. Since we have an inner spiritual body that wants to give or share and an outer physical body that wants to be filled and satisfied, there is understandably a tension and resistance between the two. Because each is enticing the other, it is a constant battle on who will win. Will the discipline and resolve of the spirit body win out or the permissiveness and apathy of the physical body? Those involved in the daily exercise of their physical bodies, know the benefits of resistance. If there is no resistance, there is no building of strong physical bodies. The same is true in the spiritual field. If there is no resistance to evil, there is no building of strong spiritual bodies! When temptations come, if the spirit body does not hold the line, the physical body does not have power within itself, to resist the pressures at hand. By learning to act for ourselves and not be acted upon by what others do, we become spiritually strong and self motivated in promoting good and resisting evil. Like the pattern of opposites within the atom, the right side then becomes the strong force that governs the whole body, not the flesh that is always wanting to be filled and satisfied.

***How Do We Record Our Answers to Life's Daily Tests*?**

We signify with our physical bodies, the path we have chosen to follow. Signify means to sign, to announce, declare, and proclaim. When we put our signatures on a piece of paper or a check, it means it comes from us because we signed it personally. The same is true of the answers we give to life's daily tests. When we signify with our bodies, we are the ones who have determined what we will or will not do. Should we obey God's laws, this obedience would have a positive effect upon our bodies because the positive energy (+) it generates would be beneficial.

On the other hand, should we disobey God's laws it would have a negative effect upon our bodies because the excess negative energy (–) generated would be detrimental. Through this process of doing, we create our own balance or imbalance. Truth cannot be hid, for we are the ones carrying out the deed. We in essence, are announcing, proclaiming, and declaring what type of person we want to become. No one can escape these truths and no one can escape their consequences, for our obedience or disobedience is being recorded within our elements. Doing the deed or repenting of the deed locks our life's history within us.

Some say the elements are evil; hence, man is depraved and cannot change his evil state. This is untrue, for it would then release us from the responsibility of our own actions. The elements are not evil, but what we choose to do with them can be either good or evil They have been organized for a special purpose and constitute a holy place, even a temple that is most sacred:

> "The *elements* are the tabernacle of God; yea, man is the tabernacle of God, even temples; and whatsoever temple is defiled, God shall destroy that temple." (Doctrine & Covenants, 93:35, emphasis added)

Our Choices Determine Who We Eventually Become

We daily experience the consequences of opposite choices. Yet some say there is no such thing as good and evil, right and wrong. They believe in "situation ethics"—that what is right or wrong depends on who is looking at the situation or which side one is on. These are false words, for both good and evil and right and wrong are as much a reality as the light of the sun and the darkness of the night. To say otherwise would be denying our very existence:

> "Wo unto them that call evil good and good evil, that put darkness for light, and light for darkness, that put bitter for sweet and sweet for bitter!" (Book of Mormon, 2 Nephi 15:20)

Some choices are of little consequence while others give cause for concern. But choices, whether big or small, determine whom we eventually become. When an unpleasant situation arises, will we choose to be pleasant or angry, peaceful or agitated, forgiving or vengeful? When we are put in a situation of upholding our honor, when it seems that no one will ever know, will we choose to be honest or dishonest, moral or immoral? When our integrity is being challenged, will we choose to be true or false, condescending or courageous? That which we choose to embrace is of utmost importance, for "*whatsoever a man soweth, that shall he also reap*." (Bible, Galatians 6:7) The act of sowing a seed seems of little consequence, but when the harvest comes in and we see what we have nurtured, we want to be pleased with the "crop."

Sometimes we make wrong choices. This is understandable, for even though we are in the process of becoming wise, this does not mean we are always wise. Learning from our mistakes is part of the process. Sometimes we are presented, not with obvious choices of good and evil, but with small, improper indiscretions. An accumulation of small, seemingly innocent indiscretions can eventually put us on the wrong path. This is why relying on our own wisdom can be a little risky. Through prayer, we can access the wisdom of our Heavenly Father. He will send the Holy Ghost, who relays His truths, to teach our spirit the safe path to follow. Our spirit can then direct our physical body to carry out the deed.

Latter-day prophet David O. McKay once said: "*The greatest battles you will ever fight will be in the recesses of your own mind.*" As a man thinketh, so is he, for in time, we play out our thoughts with actions. In this way, justice cannot be thwarted, since we create our own consequences, whether they are good or bad. It is a truism that truth cannot be hid! For in the end, our elements will reveal who we really are.

– SECTION THREE – THE LAW OF WITNESSES –

Jesus Christ: God's Greatest Witness

Up to this point, the focus has been mainly tied to the physical realm, where evidence supporting the Pawlowski Atom and the *law of opposites* has been found. The focus will now shift mainly to the spiritual realm, where a different kind of evidence supporting opposites can be found. This time the feelings of the heart, rather than the logic of the mind will be the dominant factor. Since one witness cannot stand alone in testifying of a law that takes two, the spiritual witness will be inseparably bound to the physical witness when evidence supporting the law of opposites is found.

Who Are The Witnesses of the Spiritual Realm?

They are the prophets who have spoken for Heavenly Father and His Son, Jesus Christ. Throughout the ages, they are the ones who have testified that the work of the spiritual realm is not only relevant and true, but of great importance in our lives. From them we learned that these two Great Beings created heaven and earth. The *words* that they used in describing these two different realms had to be inspired, for who else, but the ones who had created them, knew the earth was so coarse and heavy it challenged the five senses, while heaven, so pure and light, it could not be seen or physically touched? Through the whisperings of the Spirit, they were given evidences of an *unseen* spiritual world that blended with the evidences of our physical world. I have confidence in these inspired men and their use of words so I will continue to use the scriptures they brought forth in my efforts to blend the spiritual and physical worlds into one.

God's Greatest Witness

God's greatest witness has been and always will be Jesus Christ! It is understandable why his testimony has played such a vital role in man's history, for as His administrator here on earth, as well as its Creator, he has been in charge of the Father's work since the beginning. For six thousand years, he has been the world's most dominant figure; first as the premortal Jehovah and Lord of the Old Testament, then as the mortal Christ and Lord of the New Testament. Throughout the millenniums, he has been the driving force that has carried men forward and brought meaning and purpose to billions of lives. His critics will understandably not agree with such an assessment, for they think he was nothing more than an inspired teacher, or at best, like a prophet of old. But his presence cannot be so easily dismissed, for at the end time, when his work of redemption is finished and his judgments rendered, every knee shall bow and every tongue confess that he is the Savior and Redeemer of the world. By then, his power and authority will be obvious to all.

While Jesus was on earth, he was known as the great lawgiver, rabbi and teacher of the law. He taught the gospel with conviction and steadfastness, never doubting his authority to teach such truths. His words were bold and unwavering; for he knew they would be sustained. His language was interspersed with such legal terms as: "After the *trial* of your faith." "Ye will be *judged*." "The truth shall make you *free*." "The meek shall *inherit* the earth." "I am your *advocate*." "Thy sins are *forgiven*."

Why were the words that Jesus spoke couched in legal terminology? It was to convey the Father's message that all mankind will be judged by laws that are legal and binding and that at some future day, all must account for how well they lived these laws. Since it is in the courts that one's guilt or innocence is determined and one cannot be judged guilty except witnesses condemn him; his words reaffirm what the prophets have said in the past. They have outlined the terms of his impending judgment, as well as explained how mercy and forgiveness will play an equal, if not greater part. As evidence that this is true, try putting the following words in the same legal context as those above and you will see that Jesus is telling us that mercy and redemption will come to those who repent, but misery and bondage to those who will not:

> Appeal, bondage, condemn, confess, convince, corrupt, covenant, decree, exhort, guilty, innocent, inflict, intent, justify, honor, lawful, mediator, merciful, obey, pardon, punish, plead, promise, redeem, reward, repent, restitution, reconcile, swear, testimony, transgress, and witness.

We Are Compiling Our Own Court Case Within Us

The Savior's use of words is a witness that we are compiling our own court case within our bodies. We cannot sin a little, lie a little, or cheat a little and then expect at the time of judgment to receive a few stripes and all will be well. Disobedience leaves the elements in a weakened and unbalanced state. Unless we repent, the breaking of God's laws will bring consequences that are eternal and hard to bear. Christ is quoted in the following as saying that our actions, both the good and the evil, are being written in the book of our body and our spirit:

> No good deed remains unwritten before God, not from the beginning of the world. For from your kings and your governors you may wait in vain for your reward, but never do your good deeds want their reward from God.
>
> You may escape the laws made by kings, but the laws of your God, these may none of the sons of men escape. And when you come before the face of God, the devils of Satan bear witness against you with your deed, and God sees your sins written in the book of your body and your spirit and is sad in his heart. (*Essene Gospel of John, XVII*, "Christ and the Patriarchs," by Marcus Von Wellnitz, pg.172)

Many foolishly believe that what we do here on earth has no bearing on what we ultimately become. The truth is, that unless we discipline our behavior to conform to God's laws, the end product of our life will be unacceptable to Him. We are here to learn to live His laws. When we sin, He will truly forgive us if our repentance is sincere. But we are the ones who must stop the sin and remove its residue from our bodies. Why would the Savior say, "You must be found on the right hand of God," if no physical involvement was expected? And to be "born again." Is this simply a spiritual utterance? When he says, "You must become as a little child, and be tried in the furnace of affliction," it denotes physical action on our part. When he said, "Little children are whole from the beginning," was he expressing both a spiritual and physical truth? These unusual scriptures give evidence that the spiritual and physical body must work in concert if their union is to be perfected.

Jesus Christ is the Creator and Savior of the world, of this I testify by the spirit. But for those readers needing physical evidence of this reality, this next section has been written so that you might know how truth lies not only within the scriptures, but within the elements as well. The Bible, the Book of Mormon, the Doctrine and Covenants, and the Pearl of Great Price are truly the words of Christ. They are inspired spiritual witnesses. When paired with an inspired physical witness, truth is enlarged even more.

Yes, the spiritual and the physical do go together, for they are a separate part of the same system. Evidence will be presented in the remaining portions of this book to show how God intends that they work together as one.

WE MUST BE FOUND ON THE RIGHT HAND OF THE FATHER

What does this section tell us about the atom? The positive, right side governs and gives life to the left: all life forces are moving from right to left: the right side builds, the left side takes away: the right side "acts," the left side is "acted upon."

The scriptures are replete with verses that pertain to the right and left hand. Some relate to power: "*Thy right hand, O Lord, is become glorious in power,*" (Bible, Exodus 15:6) while others, relate to directional positioning: "*He is at my right hand.*" (Bible, Acts 2:25)

These two scriptures give evidence that power and directional positioning go hand in hand, with one testifying of the spiritual realm and the other testifying of the physical realm. But still, why do some scriptures say we will be blessed only if we are found on the right hand? Why not the left?

> "Then shall the King say unto them on his right hand, Come ye blessed of my Father, inherit the kingdom prepared for you from the foundation of the world." (Bible, Matthew 25:34)

Establishing Directional Positioning and Power

We know we must believe on the Savior and do the things he would do, but why equate righteous behavior with the right hand? How was this directional positioning established in the first place? I believe the answers are found in the way matter has been structured, for the Pawlowski Atom gives evidence that underlying the order and complexity of the universe is a simplicity of opposites that is remarkable. This opposite structure within the atom creates an east/west–north/south–right/left orientation that gives us the ability to create power, as well as establish directional positioning.

What evidence is there that power is being created? The illustration on page 47 clearly demonstrates that the atoms are self-generating units of power. Probability/chance has no part in the creation process, since the electrons in each octet are spinning opposite to a counterpart in the opposite octet, either up or down, right or left in two different fields of energy, as well as alternating from octet to octet. It is what we do with this power and energy within our elements that determines whether we are found on the right or left hand of God.

This same opposite orientation determines our directional positioning. As evidence that this is true, how would we know where we were at and at what given time, if these sign posts were not embodied within matter? Throughout history the North Star has helped man get his directional bearings. Sailors lost at sea or explorers unable to find their way back could determine their location by finding this bright star in the heavens. Once they found it, they knew that north would be to their front, south to their back, east to their right and west to their left. This knowledge gave them the ability to return home. A compass, with its free moving needle indicating the direction of the earth's north-south magnetic fields, determines this same directional positioning because matter is reflecting the structure of the atoms.

The following illustrates that our measuring devices are also following this same basic pattern of opposites, and that the positive right side is the stronger part that moves ahead of the left, whether in time, space, or motion.

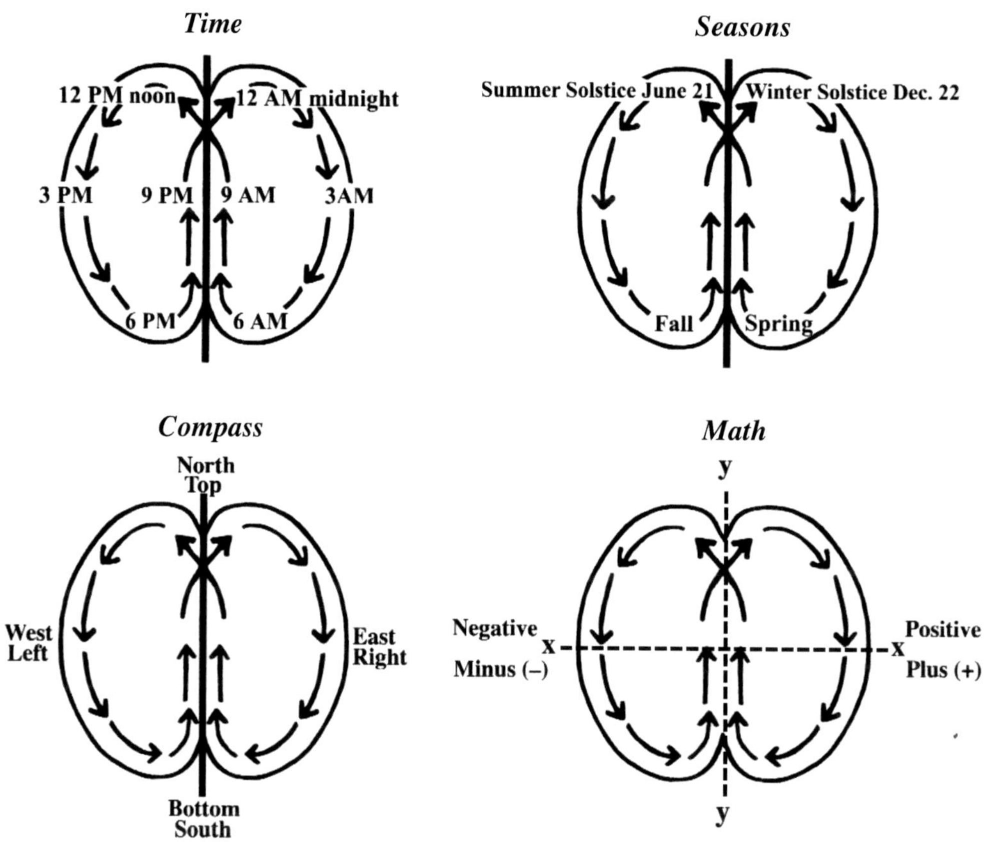

Time is measured with the twelve o'clock hour at the top, six o'clock at the bottom, three o'clock to the right, and nine o'clock to the left, with two revolutions needed to complete the AM/PM, day/night cycle. This going around twice gives evidence that the Pawlowski Atom is correct, for it, too, has the flow of energy moving in a figure eight path, first to the right, building up, and then to the left, winding down. It, too, reflects the expansion yet contraction of the Pawlowski model with time going out, then coming in. Even the seasons testify of this order, for they, too, are divided into four time intervals: winter, spring, summer, and fall. During one half of the year, (winter to summer) the days grow longer (like the positive that builds on the right side of the atom). At a precise moment, there is a flip and the seasons reverse into an opposite field, (summer to winter) with the days getting shorter (like the negative on the left that take away). Like the Pawlowski atom, seasons, too, go out and then in.

Math also uses this same right side/left side structure to get its directional bearings. Graphs that have an x and y axis—y going north/south—x going east/west—have been used for centuries to measure the value of numbers. Those on the right side of the graph are the positive, plus (+) numbers that build up in value, whereas, those on the left are the negative, minus (–) numbers that take away value. Since all the atoms have been organized with this two-part system, this supports the premise that all things are being governed by the law of opposites.

As further evidence that the right side has more energy then the left, why do calendars need extras days added every few years to bring time into alignment with the heavenly stars? I believe the extra thrust coming from the "slider electrons" in the upper right quadrant keep the right side moving ahead of the left, because it is the stronger part. If the "sliders" were not there, the electrons would be almost evenly divided with 23 on the right rays–21 on the left rays–22 on the right, inner shells–20 on the left, inner shells. It is these six "slider electrons" that create this balance, but it is also these six "slider electrons" that disrupts it. Because of this disruption, additional days must to be added to keep time in alignment.

The Polarity Barrier and the Figure-Eight Electron Path Unifies Matter

The polarity barrier and figure-eight electron path cause the electrons to work together as one. They position the electrons so that a magnetic attraction holds things together. Although each is invisible to the naked eye, they create a dipole feature that causes matter to build and expand into opposite energy fields, like the earth's magnetic fields. Since all things are made up of atoms, isn't it logical that the earth's orientation would come from the atoms that make up the earth, following the same laws? Albert Einstein envisioned the atoms with a similar two-part system. This opposite orientation creates a positive field that builds and a negative field that takes away, or in mathematical terms—a plus (+) and a minus (–). I believe this north/south, east/west structure that is established within the atoms is what causes the up/down, go/stop, plus/minus signals that bring power to our modern computers, electronic equipment, fiber optic communications, etc.

Einstein also believed that space and time are not separate entities, but a continuum, meaning space-time flows continually. The Pawlowski Atom verifies this to be true, for *like a compass, it gives our position, yet like a clock, it gives our time.* Because everything is following the same basic law of opposites that God has established within the elements, we are able to find our place in time, space, and motion. With this new knowledge, we can now find our relationship to Him.

The Right Builds/The Left Takes Away – The Right Acts/The Left is Acted Upon

We have the same directional positioning within our two bodies as we find within the elements. They, too, are found in two different energy fields, with one body having the potential to build (+), while the other to take away (–). Just like within the atom, the positive spirit body has the freedom to "act" (+) while the negative (–) physical body is restricted and has to be "acted upon." They, too, have been made to work together as one, just as heaven and earth have been made to work together as one. When the Great Creator calls them, they, too, stand up together:

> "Mine hand also hath laid the foundation of the earth, and my right hand hath spanned the heavens: when I call unto them, they stand up together." (Bible, Isaiah 48:13)

When our spiritual and physical bodies "stand up together," I believe a magnetic attraction pulls them together and holds them in place. With one being a positive entity, while the other a negative entity, they are able to produce maximum power, magnetism, energy, and light. But the right side is the initiator. It brings power to the left. The right gives, the left takes; the right is the source of power, the left is the receiver of it. This next scripture testifies of God's power coming from the right hand:

> "...Has taken his power on the right hand of his glory, and now reigneth in the heavens." (Doctrine and Covenants 49:6)

Our ability to reason and think and do comes from the intelligence embodied within our spirit body that is designed to act. It is the controlling part that brings reason to our being. It brings life because it is the activating part that sets things in motion: it is the spark of life. We have a mind and a brain, but the brain, which is in the physical body, must be activated by the mind, which is in the spiritual body. As evidence that this is true, when we die the brain is still there, but of no value because the spirit body with the mind is not there to bring it life.

Does this help in understanding why the Savior said we must be found on his right hand? It is the right side of his creations that brings life to what otherwise would have remained "as dead." It is the part that governs the left. Does RIGHTeousness take on new meaning? Besides being a directional word, it denotes good behavior. Did you know that LEFT and sinister mean the same thing? But as I said before, do not be misled into thinking left is necessarily bad. Both are needed for power to be created. It is only when our lives become too engrossed in the physical, that the left becomes harmful. Spiritually speaking, there is no such thing as becoming too perfect or too good, but we can become too evil. We can be filled with so much negative energy, that it subtracts (–) and drains the positive energy, which inherently wants to build (+). Without a strong, spiritual right side we would be found on the left hand, which has no power:

> "And the righteous shall be gathered on my right hand unto eternal life; and the wicked on my left hand will I be ashamed to own before the father. Wherefore I will

say unto them—depart from me, ye cursed into everlasting fire, prepared for the devil and his angels. And now, behold, I say unto you, never at any time have I declared from mine own mouth that they should return, *for where I am they cannot come, for they have no power*." (Doctrine and Covenants 29:27-29, emphasis added)

Jesus Christ is the Son of Righteousness

Righteousness, as used in the scriptures, is equivalent to holiness and conformity of life to divine law. Jesus Christ is called the Son of Righteousness, so we know that our Heavenly Father embodies all the attributes of righteousness. A person then, who is righteous, is God-like. He has integrity, honesty, and uprightness, and embodies all that is good. His right, spiritual, side governs all his actions, not his left, physical, side that is always wanting to be filled and satisfied.

Jesus Christ is the fountain of all righteousness and all good gifts come from him:

> "Behold, I will show unto the Gentiles their weakness and I will show unto them that faith, hope and charity bringeth unto me—the fountain of all *righteousness*." (Book of Mormon, Ether 12:28, emphasis added)

How can righteousness be obtained?

> "And blessed are all they that do hunger and thirst after *righteousness*, for they shall be filled with the Holy Ghost." (Book of Mormon, 3 Nehpi 12:6, emphasis added)

> "Angels speak by the power of the Holy Ghost; wherefore, they speak the words of Christ....for behold, the words of Christ will tell you all things what you should do." (Book of Mormon, 2 Nephi 32:3)

> "Therefore they did watch over their people, and did nourish them with things pertaining to *righteousness*." (Book of Mormon, Mosiah 23:18, emphasis added)

Jesus Christ is the one who gives us directional positioning and the power to move forward. He is the one who can heal us:

> "But unto you that fear my name, shall the Son of *Righteousness* arise with healing in his wings..." (Book of Mormon, 3 Nephi 25:2, emphasis added)

> "...there is no other way or means whereby man can be saved, only in and through Christ. Behold, he is the light and life of the world. Behold, he is the word of truth and *righteousness*." (Book of Mormon, Alma 38:9, emphasis added)

Natural Man is an Enemy to God

It is understandable why the Savior would say the natural man is an enemy to God. God's nature is like the positive elements that give or share. Because our physical bodies are made of predominately negative elements that want to be filled and satisfied, it is only natural to be concerned with our own welfare, rather than the welfare of others. We seek for things that gratify our flesh and make us feel good. With this natural, wanting to receive quality built within our physical elements, our Heavenly Father knew that all would sin and fall short of His glory. This is why repentance is so important. We must repent and bring our physical desires under subjection to our spiritual desires, or ultimately we will find ourselves on the left hand of God. It is really quite simple. *The right spiritual side must govern the whole body; otherwise we will be found on the left physical side, which has no power.*

The following scriptures testify of the positive aspects of the right hand relationship (emphasis added):

> "For behold, in my name are they called; and if they know me they shall come forth, and *shall have a place eternally at my right hand.*" (Book of Mormon, Mosiah 26:24)
>
> "Therefore being *by the right hand of God exalted.*" (Bible, Acts 2:33)
>
> "For the names of the righteous shall be written in the book of life, and unto them *will I grant an inheritance at my right hand.* And now, my brethren, what have ye to say against this? I say unto you, if ye speak against it, it matters not, for the word of God must be fulfilled." (Book of Mormon, Alma 5:58)

These next scriptures testify of the negative aspects of the left-hand relationship (emphasis added):

> "Then shall he say also unto them *on the left hand*, Depart from me, ye cursed, into everlasting fire, prepared for the devil and his angels." (Bible, Matthew 25:41)
>
> "Wherefore, I revoke not the judgments which I shall pass, but woes shall go forth, weeping and wailing and gnashing of teeth, yea, *to those who are found on my left hand.*" (Doctrine and Covenants 19:5)
>
> "And now it shall come to pass, that whosoever shall *not* take upon him the name of Christ must be called by some other name; therefore, *he findeth himself on the left hand of God.*" (Book of Mormon, Mosiah 5:10)

Indeed we see the power of God coming from His right hand. Since the positive, right, spirit body is what keeps the negative, left, physical body from becoming excessive, it is imperative that we live within the bounds He has set. Should we do otherwise, and not repent of our sins, undesirable consequences will emerge and we will be found on His left hand.

Where is the law that establishes our position in the universe? It is within the elements, it is everywhere! This right/hand versus left/ hand positioning could not have been established, had it been otherwise. The wisdom that this knowledge imparts makes this next scripture even more meaningful: "*A wise man's heart is at his right hand: but a fool's heart is at his left.*"(Bible, Ecclesiastes 10:2)

COMPLEMENTARY OPPOSITES: THE MALE AND THE FEMALE

What does this tell us about the atom? Positive and negative forces are of equal importance, but since positive is the stronger of the two, it governs and moves things forward.

In this day of confusion and soaring divorce rates, some may find it hard to believe that the male and female are complementary opposites. But they are and always will be. It is our declining social structure that has driven a wedge between the two. God has created men and women so they will work together as complementary opposites, not undesirable opposites. They have different roles and have been made differently, in order to fulfill His purposes and plan for mankind. Of course He wants all of His children to be treated with equal respect and kindness, but this equality does not mean men cannot oversee the whole or that "thy desire shall be to thy husband." (Genesis 3:16) He is not saying He esteems Adam over Eve, but rather is giving the order upon which these two opposites must operate. This next scripture clarifies this relationship:

> "*God said, it is not good that man should be alone*; I will make him an help meet for him...And the Lord God caused a deep sleep to fall upon Adam... and he took one of his ribs, and closed up the flesh instead thereof; And the rib, which the Lord God had taken from man, made he woman, and brought her unto the man. And Adam said, This is now bone of my bones, and flesh of my flesh; she shall be called Woman, because she was taken out of Man. *Therefore shall a man leave his father and his mother; and shall cleave unto his wife: and they shall be one flesh.*" (Bible, Genesis 2:18, 21-24, emphasis added)

This unity of opposites is just as we find in the Pawlowski Atom. With this as our guide, it is obvious that if things are to be created, *positive and negative forces must stand side by side*, opposite one another, for *when standing alone, they create nothing*. The electron infinity path reveals this exacting order of creation. The first electron is seated to the right of the polarity line, the second to the left, with each electron thereafter building upon that which came before, but always moving from a positive, giving and sharing energy field to a negative, receiving energy field. *The male is in the first position* because he is the "giving and sharing" entity that bestows life: whereas, the *female is in the second position* because she is the "receiving" entity that receives life. He begets, she creates. He oversees the whole because he is the strong force that moves things forward. With his larger structure and physical strength, he was foreordained to be the patriarch, protector, and provider of the family. This does not make the woman less important, for she is an equal partner, only in a different position of influence. With her smaller frame, she creates the physical bodies that house the spirit bodies, all the while serving as a helpmate and support to her husband. He presides, provides, and protects; she creates, cultivates, and cherishes.

It is interesting that within the atom, the size of the electrons to the right of the polarity are larger in circumference than those to the left. This size difference is what we find between men and women. Could this possibly be caused from the restriction put on the negative left side when the electron path goes over the polarity and magnetism occurs?

The Importance of Woman's Role

Even though the father's role is to oversee the whole structure of the family unit, in my mind, the mother's role is of equal, if not greater importance, for she has such an impact on her children's lives. Evidence of this was shown when women in their childbearing years started to work outside the home. Society's problems escalated. How could it be otherwise, for there aren't enough hours in the day for her to be all things to everyone and to do all that is expected. The need for mothers to return to the home is self-evident. Children need full-time mothers during this most important time in their lives. She leaves her mark by teaching them to be God-fearing individuals who are honest, reverent, and respectful. But it is hard to instill strong moral values in children if no one is there when valuable teaching moments present themselves.

It has been said, "The hand that rocks the cradle is the hand that rules the world." The years spent in raising children go by so quickly that individual wants can be shifted to a later date. This does not mean mothers cannot have an exciting life along the way, for learning and exploring many avenues of interest can be a lifetime adventure carried out within the home. If we put first things first, God will bless our efforts.

God's commandments consistently warn us of the dangers that lie within the physical realm. Since women are the left, negative entity, she should be aware of the *excesses* that can come to this part that wants to be filled and satisfied. The word excess denotes a warning in achieving balance and equilibrium between two forces. Many women over the years have achieved what the world has to offer—fame, power, money, and worldly goods. Yet with all their successes, it is ironic how they continued having feelings of emptiness they could not satisfy. Most found that by leaving the work force, motherhood, home, and family did indeed fill the void.

I believe the feminist movement has not produced the freedom and happiness that women envisioned in the beginning because it is basically flawed. God intends that the positive male force and the negative female force, work together as one, not separately. Those involved in the movement have been trying to be both forces. They have not succeeded simply because they are not two forces, but one force. Nothing can change that fact. Society can pass laws that mandate change, but this does not mean the change will work. As long as there are those who insist that the male and female take on the same roles and be equal in all things, the consequences of broken law will continue to disrupt our lives.

THE WORD OF WISDOM: A PRINCIPLE FOUNDED UPON OPPOSITES

What does this have to do with the atom? Foods are also found in opposite energy fields. Positive alkaline foods are to be eaten freely while negative acid foods have restrictions placed on their use.

In 1833, when The Church of Jesus Christ of Latter-day Saints was in its infancy, some of the brethren around Joseph Smith, its founder, were using tobacco. He inquired of the Lord as to the correctness of the situation. The Lord answered his petition by revealing an instruction on health called the Word of Wisdom. It is found in the 89th section of the Doctrine and Covenants (D&C). It follows the same pattern of freedom yet restriction as found in the Pawlowski Atom. Since we live in a world of choices and of cause and effect, it gives us the pattern on what we should or should not be taking into our bodies. It speaks of the need to eat fresh foods and to avoid tobacco, alcohol, and "hot drinks," which have been defined as coffee, tea, and other caffeine drinks.

The need for such a directive was not as apparent then, as it is now, for smoking was not widely accepted by women and alcohol and caffeine beverages were not so readily available. Even the nutritional value of foods was of little concern since most people lived in rural settings where there was an abundant supply of fruits and vegetables to sustain their daily needs. However, the Lord knew that in the last days, conspiring men with evil designs would necessitate the need for such instruction:

> "Behold, verily, thus saith the Lord unto you: In consequence of evils and designs which do and will exist in the hearts of conspiring men in the last days, I have warned you, and forewarn you, by giving unto you this word of wisdom by revelation..." (D&C 89:4.)

The following verses in the Word of Wisdom are especially timely in this day of fast foods, junk foods, alcoholic beverages and tobacco, for they give us a sure direction on how to sustain our bodies: (Emphasis added to all)

D&C 89:2 – "The Word of Wisdom was given to man...by revelation...*showing forth the order and will of God* in the temporal salvation of all the saints in the last days."

D&C 89:5, 7-9 – "Inasmuch as *any man drinketh wine or strong drink among you, behold it is not good*, neither meet in the sight of your Father... And again, strong drinks are not for the belly, but for the washing of your bodies." *And again, tobacco is not for the body*, neither for the belly, and is not good for man, but is an herb for bruises and all sick cattle, to be used with judgment and skill. And again, hot drinks are not for the body or belly."

D&C 89:10, 11 – "And again, verily I say unto you, *all wholesome herbs God hath ordained for the constitution, nature, and use of man*. Every herb in the season thereof, and every fruit in the season thereof; all these to be used with prudence and thanksgiving."

D&C 89:12, 13 – "Yea, flesh also of beasts and of the fowls of the air, I, the Lord, have ordained for the use of man with thanksgiving; *nevertheless they are to be used sparingly*: And it is pleasing unto me that they should not be used, only in times of winter, or of cold, or famine."

D&C 89:14, 17, 15, 16 – "All grain is ordained for the use of man and of beasts, to be the staff of life...Nevertheless, wheat for man... And these hath God made for the use of man only in times of *famine and excess of hunger*...also the fruit of the vine; that which yieldeth fruit, whether in the ground or above the ground."

Positive Instruction Gives Freedom – Negative Instruction Restricts

Instructions of a positive nature are found in verses 10, 11, and 16. They encourage us to eat fresh fruit (in the season thereof) and vegetables (in the ground and above the ground) and herbs.

Instructions of a negative nature are found in verses 12 and 13. They indicate some restrictions on the eating of meat. Even though meats are to be eaten sparingly, this does not mean they are not to be eaten at all. They are to be used mainly in times of winter, cold, and famine, which of course would be when fruits and vegetables were not so readily available.

Grains (wheat) have both a positive and negative instruction. We are told in verses 14 and 17 that grain is ordained to be the *staff of life*, nevertheless, wheat is for man. Since a staff is something we lean on for support, wheat is to be the mainstay of our diet. This would especially be so during times of famine since fresh foods would not be readily available. But what is meant by *excess of hunger*? Certainly during the day when our energies are depleted and we no longer have strength, would we not experience different levels of hunger, according to our activity levels? It would seem logical that those with high-energy jobs would need more wheat than for those with low energy jobs. And would children not experience "excess of hunger" more frequently because of their high-energy output than older people with their limited activity? It would seem that *our wheat intake should reflect our ability to burn energy*. Since wheat provides more long lasting energy than fruits and vegetables, as a complex carbohydrate, it has the ability to release energy for longer periods of time.

The strongest restrictions are found in verses 5, 7–9. They warn that tobacco, alcohol, and hot drinks (coffee, tea, and cola) are harmful to the body and should not be used at all. They are members of the alcohol and alkaloid families (no relation to alkaline). *The Lord has ordained them for purposes, other than their present use*. He never intended alcohol and tobacco to be ingested by man because eventually they cause death. Helen Pawlowski called these substances "deceivers," for at first they soothe the body, then shortly thereafter, a let down occurs and the body wants more. I believe this is because they are made up of negative atoms that want to be filled and satisfied with electrons. They create a vicious cycle within us, for the more we try to satisfy the cravings they bring, the more dissatisfied we become. They are stumbling blocks in our efforts to change because the need to gratify becomes stronger than the desire to obey. Satan has taken that which the Lord ordained for other purposes and effectively uses them to deceive man. Because of their ability to kill bugs, some deceivers make excellent insecticides. Man is now using them to bring death to his body.

The Word of Wisdom Shows the Order and Will of God

His Order – In verse 2 of the 89th section it states that the Word of Wisdom shows forth the "order and will of God." What does this mean? I believe He is telling us there is a particular order even in the foods and substances we take into our bodies. Like the pattern of opposites found in the Pawlowski Atom, foods too, are found in two opposite energy fields—alkaline and acid. Those classified as alkaline are a complex form of positive, while those classified as acid are a complex form of negative. Each food is made up of elements that have stored within them a predominate amount of one or the other type energy. During the process of digestion, this energy is released.

Helen Pawlowski points out that there is an unexpected reversal in the way food tastes and their actual affect upon the body at the end of the metabolic process. They actually reverse themselves so that foods testing acid, end up leaving an alkaline ash, while foods testing alkaline, end up leaving an acid ash. Could this be evidence of the polarity and the flip, as shown in the Pawlowski Atom? Do foods actually move from a positive field to a negative field or a negative field to a positive field during the process of digestion? This seems very possible, for when placing nitrazine paper on a lemon or orange, they test acid. Yet when broken down within the body, they reverse themselves and leave an alkaline ash. Meat and eggs are the opposite. Outside the body they test alkaline, yet when going through the process of digestion, they turn acid.

As evidence that foods change when their stored up energy is being burned within the body, which foods are hard to digest, and which foods never seem to cause digestive problems? Fruits and vegetables, which test acid, yet leave an alkaline ash, are easy to digest, while meat and eggs, which test alkaline, yet leave an acid ash, take longer to break down; often leaving an uncomfortable, over acid stomach. Helen Pawlowski found further evidence on this reversal of foods:

> Proteins such as meat and eggs, when burned in the body result in a number of end products which are acidic, such as uric acid, sulfuric acid, and phosphoric acid ...*Acidity or its effect in the body are not indicated by its taste or its original acidity.* (*Tabor's Cyclopedic Medical Dictionary*, Edition 10, pg. A-17, emphasis added)

His Will – The Word of Wisdom reveals "God's will" by giving us a pattern on what He wants us to "act" or "not act upon." Keep in mind that foods and substances can have no affect upon our bodies unless we pick them up, eat them, or ingest them in some manner, for they are inert, physical objects that cannot act for themselves. They must be acted upon. The Word of Wisdom, by defining the parameters in which we must operate, is outlining what the physical body "can" or "cannot do." The instruction is simple. *Things to the right are to be used in abundance, while things to the left are to be used sparingly and within the bounds He has set.*

This right side–left side structure can be applied to God's commandments. *The positive commandments* could be called the "thou shalts," for they basically *tell our spirits what our physical bodies "can act upon."* They tell us (among other things), to love the Lord Thy God, with all our heart, mind, and strength; to honor our Fathers and Mothers; and to keep the Sabbath day holy. To obey positive commandments, we must do something about them with our physical bodies. We must be moved to good works. In this realm we have unlimited freedom to go forth on our own initiative. When obeyed, their consequences leave the one giving as well as the one receiving with more positive (+) energy.

In like manner, there are *negative commandments that tell our spirit bodies what our physical bodies should "not act" upon.* These could be called the "thou shalt nots," for they basically say we must "not act" upon another person's property (stealing), another person's wife (adultery), destroy another person's body (killing), or defame another person's good name (false witness, lying). To obey negative commandments, we must refrain from doing certain things with our physical bodies. These commandments are definite and restricting. When disobeyed, they leave the perpetrator as well as the victim with less. Like the negative elements that have empty spaces within their structure, the consequences they create leave minuses (–). It is understandable why God wants us to obey his commandments. Positive actions leave us with energy that builds while negative actions, with energy that takes away.

The Word of Wisdom is Both a Physical and Spiritual Witness

The world has long been able to dismiss spiritual witnesses for they cannot be seen nor physically proven to be true. Commandments such as "thou shalt love the Lord thy God" and "thou shalt keep the Sabbath Day holy," are irrelevant to some because they see no tangible evidence that they exist. However, just because we cannot see something, does not mean it doesn't exist. The physical world is teeming with elementary forms of matter that cannot be seen by the naked eye. Science well understands this reality as they daily experiment with the world of the small, yet having complete confidence that the unseen is there.

A physical witness cannot be so easily dismissed. Even though the Word of Wisdom refers to physical objects that are real, not abstract things of the spirit that can be rationalized away, the Lord is revealing the pattern by which all things, both seen and unseen, are being governed. He is giving evidence that as elementary matter develops into more complex entities, it continues to be bound by the same laws of freedom, yet with restriction, as found in the Pawlowski Atom. How is the Lord's "order and will" made known? *His order* is revealed by showing the same positive/negative order in foods and substances, as we see in the minute atom. *His will* is revealed by showing the boundaries in which the physical body must operate.

Is There a Life or Death Situation Within Our Elements?

Years ago, men of science began to suspect there was more than a casual relationship between the alarming rise in certain type deaths and our new modern lifestyle. They felt these deaths had something to do with what we ate and our consumption of tea, coffee, tobacco, alcohol, and cola. After years of experimenting, their feelings were verified. They found that what we take into our bodies is important and that some things cause a life or death situation within our elements. This presents an interesting parallel. Science is putting its energy into helping mankind overcome death, yet, on the other hand, religion is doing the same. But there is a difference. Science is concerned with the physical death whereas religion, the spiritual death. Both are important, but the Savior has said:

> "Wherefore, fear not even unto death; for in this world your joy is not full...Therefore, care not for the body, neither the life of the body; but care for the soul, and the life of the soul." (D&C 101:36-37)

He knows that in this life our joy is not full, therefore we should be more concerned with the life of the spirit, than the death of the physical body. This is not to say he is unconcerned with the suffering physical death brings, but rather he knows the finality of the spiritual death as opposed to the temporary nature of the physical death.

When the Lord says to stay away from certain substances and to eat some foods sparingly, he is telling us that some things bring death to the body. Even in the Garden of Eden we see eating as the act that brought death into the world. How could eating upset the elements? I believe it is because too many negative type foods cause toxins to build up within the body. Excessive amounts of acid foods convert into excessive amounts of negative energy; more than the body can handle. Helen Pawlowski felt that physical death generally occurs when the spirit body can no longer remain in a compound body whose negative energy, over the years, has exceeded its positive energy.

We must remember that the physical body is made of mainly negative elements whose atoms have empty spaces within their structure. By its very nature, it wants them "filled and satisfied" with electrons. How do we see that this need is met? We keep our physical wants within the bounds the Lord has set. If we consume more than is necessary, the balance and equilibrium in our bodies is upset, and the excess negative energy that is generated leaves us with more "needing to be filled" atoms, than our giving side can supply. In an attempt to eliminate the feelings this emptiness brings, we end up consuming more. It is like a beast devouring his own tail. The more and faster he eats, the sooner he brings about his own death. He is like the negative element that always wants to be filled and satisfied.

When the Savior said, "natural man is an enemy to God," he knew the nature of the negative element and its ability to become destructive if there were no restraints:

> "Wherefore, men are free according to the flesh; And now my sons, I would that ye should look to the great Mediator and hearken unto his great commandments;... and choose eternal life, according to the will of his Holy Spirit; *And not choose eternal death, according to the will of the flesh* and the evil which is therein, which giveth the spirit of the devil power to captivate, to bring you down to hell, that he may reign over you in his own kingdom." (Book of Mormon, 2 Nephi 2:27-29, emphasis added)

Hereditary diseases, viruses, birth defects, and accidents, alter to a great degree our physical health. But we must seriously consider that by our own choices, we either knowingly or unknowingly speed up the process of death. By willfully disregarding the words of the Savior and following the enticements of the evil one, we put excessive negative energy into our physical bodies, thus causing power to be drawn from the spirit body that is bringing us life. This is why our spirits must be strengthened. They must have the power to restrain our physical bodies from acting upon objects that cause death.

As an example of this truth, man is compounding his problems even more by using a new type of negative—hard drugs. They hasten his death because of their ability to add, in an instant, massive amounts of negatives to the body. Some call this invasion as "blowing your mind." Is it a coincidence that "acid" is the name used to describe these drugs? Drug users bombard their bodies with more negatives than the Great Creator has allowed, thus causing power to be drawn from the governing spirit that should be restraining the whole body.

Spiritual and Physical Rewards Come From Obeying the Word of Wisdom

The Word of Wisdom was given as a principle with promise. Not only should the strong and healthy follow its precepts, for even the weakest are promised rewards if they adhere to its principles: "*Adapted to the capacity of the weak and the weakest of all saints, who are or can be called saints*." (D&C 89:3)

The law of opposites is manifest in the blessings that come to those who obey the Word of Wisdom. One is physical in nature, the other spiritual:

> "...All saints who remember to keep and do these sayings...shall receive health in their navel and marrow to their bones...*And shall find wisdom and great treasures of knowledge, even hidden treasures*; And shall run and not be weary, and shall walk and not faint." (D&C, 89:18-20, emphasis added)

The spiritual promise of finding wisdom and great treasures of knowledge should give comfort to those who have followed the Lord's health instructions, only to continue suffering poor health. For even though their physical problems seem to always remain, would not the spiritual blessing be the greater prize?

God is Guiding the Affairs of Men

This knowledge on the care of the physical body will be of greatest worth if it helps us realize there is a God in Heaven guiding the affairs of men and that He speaks to men on earth through His chosen prophets. Joseph Smith was a prophet of God, and had it not been revealed to him, he too would not have known God's physical laws, which in the end, testify of His spiritual laws.

The Word of Wisdom reveals the "order and will" of God by showing that both the right and left side of His creations have been ordained for man's use, but that the left side should be used sparingly or within the bounds He has set. It stands as a witness that He knew a law of opposites was governing matter; otherwise, he would not have revealed its bounds. If we are restrained in the type of foods we eat and the things we take into our bodies, then the things we see, hear, do, and say must surely have these same restraints. Although all must die the physical death, a loving Heavenly Father has given us a Word of Wisdom so that we might live a longer, more productive, and problem free life.

STAY TO THE RIGHT, LIKE THE FIG, AND LIVE

What does this have to do with the atom? It shows the right side builds whereas the left side takes away. Also the right side brings life and sustains life.

The Word of Wisdom relates to what our physical body "ought or ought not to be acting upon." Since our spirit and physical bodies are inextricably bound together, it concerns our whole being. Its importance cannot be dismissed, for the physical body must have a degree of health, otherwise the spirit body would have nothing "to act upon." Should this part be missing, the two could not work together as one.

As we look closer at the foods the Lord has ordained "for the constitution, nature and use of man," we see why he has given such instructions. Each food creates a different condition within the body. Some produce positive energy, others negative energy. Some are short acting, others long acting. Some go through the body rapidly, others slowly. Some foods ward off disease while others make us susceptible to disease. Science has found these differences affect the long-term health of the body. Alkaline foods bring greater physical health because they produce fewer toxins in the body. They are the wholesome fruits and vegetables the Lord has said to eat in the season thereof. Some of their strong distinguishing characteristics are that they are fresh and alive, with high fiber and chlorophyll, and low in fat. They are to be used with prudence and thanksgiving; otherwise, there are no restrictions on their use.

There are other foods that must be used within the bounds the Lord has set. He specifically mentions meats are to be used sparingly, or "only in times of winter, cold, or famine," and that wheat is to be used in times of famine and *excess* of hunger. It is easy to see why meats are to be used in times of cold, winter, or famine. This is when fresh fruits and vegetables are not so readily available and when more energy is needed to heat the body. It is also easy to see why grains are to be used in times of famine. They can be stored for long periods of time since their nutritional value stays locked within the shells until cracked.

But what does the Lord mean by excess of hunger? Could it be a daily thing that occurs when our physical energies are depleted and we no longer have strength? Since he specifically states that wheat is for man, this would mean that wheat is the best food to alleviate our daily hunger because it provides more long lasting energy than fruits and vegetables. Excess of hunger would differ in various types of people. Those who are sedentary and unable to get around would not experience hunger like those working a full eight hours at hard labor. Young children with their boundless energy would also need more of wheat. Our wheat intake should reflect our activity. Some need more energy, others less. It depends upon our ability to burn energy.

The following explains why wheat is the staff of life and why it is to be eaten when experiencing "excess of hunger." We have not always known the reasons behind the Lord's health instructions, but as scientists expand their knowledge of foods, the answers seem to always surface:

> ...A complex carbohydrate contains both starch and fiber. Chemically they are chains of many sugar molecules. Complex carbohydrates are the body's best fuels for energy. They provide essential vitamins and minerals and most are low in fat. Their fiber helps the digestive system work properly...Simple carbohydrates are found in fruit. They can be defined as having one or two molecules linked together, usually sucrose and fructose. Simple carbohydrates give an initial burst of energy but soon leave the individual feeling more tired. (*Mesa Tribune*, "Weighing the Consequences," Carrie White, Aug. 17, 1988)

Wheat gives the body a feeling of satisfaction longer than other foods because its numerous sugar molecules act like time released capsules that burn slowly. This characteristic helps the body maintain a higher level of energy for a longer period of time. The fiber in the grains adds beneficial bulk to the diet as well as slows the absorption of sugar.

Science has also found that we can dramatically lower our cholesterol levels by using fats derived from the plant kingdom rather than the animal kingdom. This is most interesting since plants and vegetables are in the alkaline-positive families while meats in the acid-negative families. Eating less fat and animal proteins can also lesson our chances of cancer and heart disease. These are all evidences of the truthfulness of the Word of Wisdom. It teaches correct principles that enable our physical bodies to be sustained and strengthened, thereby allowing our two bodies to continue working together as one.

Nature's Most Alkaline Food

When listed in order of alkalinity, one study shows that the fig is the most alkaline of all fruits and vegetables (see pg. 123–124). It is so far above other foods, that *it is like a symbol telling us to stay to the right, like the fig, and live!*

We read of the exceptional healing qualities of the fig in this next scripture: "*And Isaiah said, Take a lump of figs. And they took and laid it on the boil and he recovered.*" (Bible, 2 Kings 20:7)

One leading characteristic of the fig that would cause things to heal comes from its ability to rid the body of toxins, for the higher level of our alkalinity, the fewer toxins we produce. I found further evidence of the figs exceptional healing qualities in my own refrigerator. On the label of a bottle of fig concentrate were these words: "*The fig itself is natures most perfectly balanced food*. The high alkaline and mild laxative properties help rid the body of toxins while supplying nutrients. It's also low in sodium and high in potassium." (Miss Figgy Concentrate, R.W.Knudsen and Sons, Inc., Chico, California)

I decided to see for myself how perfectly balanced the fig would be, so at that time

I opened a quart jar of Fig Pep, used some of the contents, then put it back on the shelf with the lid on loose. Every few years I would check back to see if it had spoiled. To this day, twenty-six years later, it has not. This tells me that its atoms are perfectly balanced and whole.

It is interesting that my sister Helen Pawlowski, who for the past 50 years has been eating figs in one form or another on a daily basis, has brought forth this new knowledge of the atom. Because of her regular purchase of figs or fig concentrate, the store clerks would often call her "the fig lady." Her extraordinary work in discovering the inner workings of the atom gives credence to the premise that regular use of figs will help in the enlightenment processes of the mind. The fig has within it, compounds that assist in the body's production of serotonin, a brain nerve transmitter that aids in the electrical impulses of the brain. This quality would greatly enhance man's search for enlightenment. The scriptures even say that in the last days "*...they shall sit every man under his vine and under his fig tree; and none shall make them afraid*."*(Bible, Micah 4:4)* Could this mean that men and women everywhere will receive further enlightenment? Most certainly, for God's work is not yet finished.

Can you think of anything the world needs more than enlightenment and healing? What would be more enlightening than to know the Savior created the world by organizing the elements into a beginning set pattern of opposites that brings balance and harmony to the whole, and what would be more healing than to live this law that has been established within the elements?

Nature's Most Acid Food

The yolk of the egg is nature's most acid food (see pg. 123). This is most interesting since life begins in the yolk of an egg. You would think with such a high level of acidity, life could not be reproduced. This seems to give evidence of the need for a strong magnetic connecting of positive and negative energies for life to form.

Is There a One On One Balance Among the Building Blocks of Creation?

Matter, as we understand it, is made up of more positive elements than negative (71 positive, 15 negative and 6 inert gases). This lopsided, uneven ratio among the building blocks of nature requires that creation be one of balance and equilibrium, rather than exact equal proportions. This same ratio seems to be found throughout all creation. We thrive on more positives than negative, more alkaline than acid and more spiritual than physical. We enjoy sixteen hours of day, as opposed to eight hours of night. We enjoy sweet more than bitter, quiet more than loud, and light more than dark. Could it be that we need smaller amounts of negative energy because the weight between the two are different? Could the *unrestricted* nature of the positive field be causing spirit matter to be smooth and light, while the *restricted* nature of the negative field, course and heavy? The law of opposites tells me this must be so, for when elements start forming to the left of the polarity, matter would start compressing together and become course and heavy because the magnetism between positive and negative energy is now in place. When the scriptures speak of sin (which comes from Satan) as being a heavy burden that weighs a person down, while the Savior's yoke is easy and his burden is light, it is more understandable: "*My yoke is easy and my burden is light.*" (Bible, Matthew 11:30) Things of the spirit leave a light, airy feeling whereas things of the flesh (if excessively abused) leave a heavy burdened down feeling.

Natural Alkaline/Acid Ratio Within The Body

This uneven ratio between the positive and negative is also found in foods, as shown in the following, taken from a pamphlet called, *Foods That Preserve The Alkali Reserve,* by Arthur W. Snyder. It laid the foundation of Helen Pawlowski's thinking concerning opposites. From this material she could see the consequences that would come from eating more fresh alive alkaline foods than acid foods.

Snyder said that the proper ratio between acid/alkaline foods should be four parts alkaline to one part acid, thus giving the body an alkaline reserve of three. Should it drop lower than this our health would be at risk:

> The natural ratio between alkaline and acid elements in the normal healthy body is approximately 4 to 1, giving an alkaline reserve of 3: the food intake should be in the same proportion, about 4 parts alkaline foods to 1 part acid. When the proper ratio between acid and alkaline foods is established a strong resistance and immunity to disease result. If the alkaline-acid ratio drops to 3 to 1, health is seriously menaced; a surprising number of people today are dangerously near this condition; should the alkaline-acid ratio ever drop to 2½ to 1, death would occur. *Life is possibly only the presence of an adequate positive alkaline condition of the body tissues and blood.* (emphasis added)

Acidity and Alkalinity Numbers of the Most Basic Foods

Snyder compiled a list of foods to describe their acid/alkaline values. He says, in explaining this list:

> The acid and alkali numbers given in the following food tables express the approximate potential acidity and the approximate potential reserve alkalinity. These numbers are obtained by expressing the amounts of *sulfur*, *phosphorus*, and *chlorine* present in the acid forming foods as the equivalent number of cc (cubic centimeters) of normal acid; and the amounts of *potassium*, *sodium*, *calcium*, and *magnesium*, in the alkalizing foods, as equivalent numbers of cc of normal alkali. The data obtained by the mineral analysis of foodstuffs show a definite relationship to their chemical behavior in the body, as regards their influence on the acid-base balance. (emphasis added)

Keep in mind that the first three italicized elements; sulfur, phosphorus, and chlorine are found on the left, negative side of the atom while the last four; potassium, sodium, calcium, and magnesium, are found on the right, positive side in strong positions on the first two rays. From my own experience, I found that these last four positive elements play a key role in keeping the bodies electrical system strong. (See pages 124–125.)

Snyder's List of Acid - Alkaline Foods

Acid Forming Foods—1 ounce portions (30 grams)

Food	Acid No.	Food	Acid No.
Egg Yolks	7.5	Goose	3.0
Herring (smoked)	5.5	Turkey	3.0
Oysters	4.5	Barley	3.0
Crab	4.5	Cereals (in general)	3.0
Lobster	4.0	Whole Egg	3.0
Oatmeal	3.6	Macaroni	3.0
Sardines	3.4	Spaghetti	3.0
Veal	3.3	Rice (white or brown)	2.8
Salmon	3.3	Haddock	2.5
Chicken	3.2	Walnuts	2.4
Liver	3.0	Bread	2.2
Beef	3.0	Egg White	1.7
Pork	3.0	Cheese	1.6
Bacon	3.0	Lentils	1.5

Alkali Forming Foods - 1 ounce portions (30 grams)

Food	Alkali No.	Food	Alkali No.
Figs (dried)	30.0	Potatoes	2.0
Molasses	18.0	Apricots	2.0
Olive	16.0	Pineapple	2.0
Lima Beans (dried)	12.0	Baked Beans	2.0
Soy Beans	12.0	Cherries	1.8
Apricots (dried)	9.5	Nectarines	1.8
Spinach	8.0	Cabbage	1.8
Mustard Greens	8.0	Currents (dried)	1.7
Dandelion Greens	8.0	Radishes	1.7
Beet Greens	8.0	Grapefruit	1.7
Kale	7.0	Orange	1.7
Raisins	7.0	Sauerkraut	1.7
Swiss Chard	5.0	Tomatoes	1.7
Lima Beans (fresh)	4.5	String Bean	1.6
Parsnips	3.6	Lemons	1.6
Almonds	3.6	Cauliflower	1.6
Carrots	3.5	Peaches	1.5
Beets	3.5	Mushrooms	1.2
Dates	3.0	Apples	1.1
Rutabaga	2.5	Pears	1.1
Rhubarb	2.5	Bananas	1.1

	Allkali No.		Alkali No.
Celery	2.5	Grapes	1.0
Cucumber	2.5	Watermelon	1.0
Cantaloupe	2.2	Squash	1.0
Lettuce	2.2	Buttermilk	0.7
Parsley	2.2	Whole milk	0.5
Pomegranate	2.0	Green Pea	0.3
Watercress	2.0	Asparagus	0.2
Coconut	2.0	Ice Cream	0.1

Alkalinity numbers of the more important *juices* (per pint 480 grams)

	Alkali No.		Alkali No.
Fig	200	Pineapple	16
Beet Greens	64	Pomegranate	16
Dandelions	64	Watercress	16
Spinach	64	Cabbage	14
Turnip Leaf	64	Grapefruit	13
Parsnip	28	Orange	13
Carrot	28	Tomato	13
Beet	28	Lemon	13
Celery	20	Apple	9
Apricot	16	Grape	8
Coconut	16	Watermelon	8

From the above, we see that positive alkaline foods are mainly the fruit, vegetable, and bean families, while negative acid foods are mainly the meat and grain families. Both are for the constitution, nature and use of man, but acid foods are to be used sparingly or within the bounds the Lord has set. As added evidence that this is true, Dr. Ragnar Berg, an authority on the subject, said that we should be eating a ratio of 80 alkali-forming foods to 20 acid forming foods (*The How To Herb Book*, Keith and Gordon, pg. 162). This is especially interesting, since the numbers 2 and 8 play such an important part in building the Pawlowski Atom.

Perhaps the reason alkaline foods are so beneficial is that they play an important role in keeping the body's electrical system working properly. The following quote gives further evidence that the body is an electrical unit and that certain positive alkaline elements are needed for the system to work properly. It also gives compelling evidence that positive energy is what moves things forward:

> Researchers know the body normally has tiny electrical currents flowing throughout. These currents are created by the flow of electrically charged particles called ions. *In the body, ions are usually of calcium, potassium, sodium, or magnesium.* (Authors comment: These are all strong positives found on the first two rays).

> They flow in and out of the body's cells.... These currents start at the moment of conception and seem to guide development and differentiation in the cells of the growing fetus. Scientists like Dr. Bassett (director of Columbia Universities Orthopedic Laboratory in New York City) believe these currents act to control many chemical events in the body. Messages they send to cells influence their growth and function. (*Prevention*, "Electricity's Healing Potential," Nov. 1985, pg.73, emphasis added)

Taking Control Of Our Lives

More evidence linking science to the Word of Wisdom comes from the American Cancer Society. Their pamphlet, *Taking Control*, lists ten steps that lead to a healthier life and reduced cancer risk. The foods they recommend most are the highly alkaline foods, which is another witness that the right side brings life:

> Today, scientists think most cancers may be related to lifestyle and environment—what you eat, drink, if you smoke and where you work and play. So the good news is you can help reduce your own cancer risk by taking control of things in your daily life.
>
> We have five protective factors for you to add to your life and five risk factors to subtract.

What are the Protective Factors?

1. *Eat more cabbage-family vegetables*: Important studies show these vegetables (also known as cruciferous) appear to protect you against colorectal, stomach, and respiratory cancers. They include broccoli, cauliflower, Brussels sprouts, all cabbages, and kale. (*These are the right side, alkaline foods.)
2. *Add more high-fiber foods*: A high fiber diet may protect you against colon cancer. Fiber occurs in whole grains, fruits and vegetables including peaches, strawberries, potatoes, spinach, tomatoes, wheat and bran cereals, rice, popcorn, whole wheat bread. (*These foods are both acid and alkaline, but the grains, which are acid, do not come from an animal source.)
3. *Choose foods with Vitamin A*: It may help protect you against cancer of the esophagus, larynx, and lung. Fresh foods with beta-carotene like carrots, peaches, apricots, squash, and broccoli are the best source, not vitamin pills. (*These are alkaline foods.)
4. *Do the same for Vitamin C*: This vitamin may help protect you against cancers of the esophagus and stomach. You'll find it naturally in lots of fresh fruits and vegetables like grapefruit, cantaloupe, oranges, strawberries, red and green peppers, broccoli, tomatoes. (*These are alkaline foods.)
5. *Add weight control*: Obesity is linked to cancers of the uterus, gallbladder, breast, and colon. Exercise and lower calorie intake help you avoid gaining a lot of weight. Walking is ideal exercise for most people, and primes you for other sports. Check with your physician before strenuous activity or a special diet.

What Are The Risks Factors?

6. *Trim fat from your diet*: A high-fat diet increases your risk of breast, colon, and prostate cancer. Fat-loaded calories mean a weight gain for you, especially if you don't exercise. Cut overall fat intake by eating lean meat, fish, skinned poultry, low-fat dairy products. Avoid pastries and candies.
7. *Subtract salt-cured, smoked, nitrate-cured foods*: Cancers of the esophagus and stomach are common in countries where these foods are eaten in large quantities. Choose bacon, ham, hot dogs or salt-cured fish only occasionally, if you like them a lot.
8. *Stop cigarette smoking*: Smoking is the biggest cancer risk factor of all—the main cause of lung cancer and 30% of all cancers. Smoking at home means more respiratory and allergic ailments for children. Pregnant women who smoke harm their babies. Chewing tobaccos are harmful, too, as risks for mouth and throat cancers. Pick a quit day now and call for help.
9. *Go easy on alcohol*: If you drink a lot, your risk of liver cancer increases. Smoking and drinking alcohol greatly increases risk of cancers of the mouth, throat, larynx, and esophagus. If you do drink alcohol, be moderate in your intake. (*Abstinence is the only way because alcohol is addicting).
10. *Respect the sun's rays*: Too much sun causes skin cancer and other damage to your skin. Protect yourself with sunscreen—at least #15, wear long sleeves and a hat, especially during midday hour—11 a.m. to 3 p.m. Don't use indoor sun lamps, tanning parlors, or pills. If you see changes in a mole or a sore that does not heal, see your physician.

> Not all cancer risk factors are equal. The most important risk is tobacco. And some risk factors, when combined, greatly increase risk, smoking and heavy drinking, for instance. However, you can't control all risk factors like family health history or your heredity. (End of quote used from *Taking Control* pamphlet.)

*All starred notes are this authors comment.

We Need More Positive Energy Than Negative Energy

This information from the American Cancer Society and Snyder's study on foods supports one another. Each shows the desirability of eating positive alkaline foods over negative acid foods. Each gives evidence that there is not a one on one ratio between the positive and the negative, but one of balance and equilibrium and that the right side governs, since positive (+) is stronger than negative (–). It must be remembered that the Word of Wisdom said the same thing 170 years ago, but without the benefit of the scientific knowledge we now have. Truth is truth, and whether found in the spiritual or physical realm, it comes from God. Can anyone doubt that too much negative energy within the body causes problems? This same balance between positive and negative energies is found throughout so many areas of creation that it would be well to use caution whenever adding negative energy to our bodies in whatever form.

Another compelling argument for positives recently surfaced. Many doctors feel that free radicals are the main cause of aging. And what are free radicals? They are molecules looking for electrons. Since a molecule is a more developed form of an atom, this would mean they are negative molecules since a negative always lacks one or more electrons. Free radicals steal electrons from other molecules in order to achieve balance and equilibrium:

> Free radicals are as wild—and promiscuous—as their name implies. Like a street gang run amuck, they roam the body as molecular thieves. *These highly reactive molecules attack other molecules, forcing them to surrender one of their electrons.* "Because free radicals have an unpaired electron, they are chemically very reactive," explains biochemist Denham Harmon, MD., Ph.D., of the University of Nebraska Medical Center, who proposed the free-radical theory of aging in 1955, when few scientists thought free radicals were linked to disease, let alone aging. "They can start a chain reaction that does real damage to the body." (*Walking Magazine*, Nov/Dec. 1990, "When 'rads' are for real," Carlson and Seiden, emphasis added)

Like the Minute Atom, Our Bodies are Striving For Equilibrium and Balance

The Pawlowski Atom gives convincing evidence that balance and equilibrium are an integral part of the universe because matter has been organized that way. When one understands this model and how the atoms and elements have been formed, one can see why balance is so fundamental to God's plan and why the desire for stability and order is inherent within each individual. With two energy fields at work within the body, it demands there be a balance between the two. He has given us the Word of Wisdom to show how this balance can be maintained. To know that alkaline foods (+) are to be eaten in abundance and acid foods (–) sparingly helps us understand why undesirable consequences emerge if this balance is disrupted. And to know that a positive force (+) must lead the way in all of God's creations helps us realize the importance of having a strong spirit to restrain our physical wants.

Too much of the left is the problem of our day. With its built in desire of wanting to be filled and satisfied, God has put restrictions and bounds on its use. But man continues to act as if there were no restraints so man continues to suffer its consequences. *Can you see why it is appropriate to say, "Stay to the right, like the fig and live?"* The right side has the ability to build and sustain life while the left side, to destroy and take it away!

ATONEMENT/JUDGMENT: MERCY OFFERED; YET THE LAW SUSTAINED

What does this have to do with the atom? When Gods laws are broken, He provides mercy, like the positive element that wants to give and share. He also executes judgment, like the negative element that wants to be filled and satisfied.

During this mortal life, we all at one time or another do things that are contrary to the nature of God (a minus –). Some misdeeds could be attributed to moral laxness while others to downright rebellion. Some are of a grievous nature, others are not so offensive. When this happens, we are to a greater or lesser degree sinning, for we are going against the laws that God has established within the elements. Since these laws bind all creation, He works through the same binding laws. Should they be broken, He requires that penalties be paid, for with every broken law there is a consequence that must be dealt with. But, what if during the process of breaking the law, we realize the error of our ways and want to turn from sin? Is there a way we can be forgiven or are we forever burdened with the consequence of our follies?

There is a way to be forgiven, a way in which we are able to start anew. Our Heavenly Father has given us a "Plan of Salvation." Since salvation means the act of saving or delivering, this plan is just what it says it is—"a plan that saves" (a plus +). But I might ask, saved from what? From death and sin, these two most dreaded words that all must come to know! We have all experienced the grief that comes from these two sources, sometimes even questioning their value in our lives. But since each is a reflection of laws being broken, their purpose in the great eternal scheme of things is to teach us how to become wise and all knowing like our Heavenly Father. They teach us that opposition is governing all things and that obedience brings happiness not sadness, life not death, sweetness not bitterness.

The Need For a Savior

Heavenly Father knew we would in varying degrees break the law while learning to obey it. He therefore provided a way in which the errors we make in this mortal life can be corrected. Even though He cannot look upon sin with the least degree of allowance, He is merciful and wants to help us in our need. To make up for our deficiencies, He provided a *Savior* to redeem and deliver us from the consequences of broken laws. Since a Savior means one who saves, rescues, or delivers; and redeem means to buy, pay off, and to clear payment, this deliverer will liberate us from death and release us from the ill effects of sin. Who is this Savior who will rescue us from broken laws? It is Jesus Christ, God's Only Begotten Son in the flesh! He who was the great Jehovah of old and the long awaited Messiah of Israel! In the pre-existence he created the world, therefore, as its central figure, will at the end time, be our judge.

When Jesus created this beautiful world, the intelligence's from the small to the very great gave him their allegiance, knowing full well that he would uphold the laws by which all things were being governed. They fill the immensity of space and they answer to his call:

> "Behold, great and marvelous are the works of the Lord...For behold, by the power of his word man came upon the face of the earth, which earth was created by the power of his word." (Book of Mormon, Jacob 4:8-9)

Because Jesus had power over the elements, he was able to walk on the waters and calm the troubled storm. Yet with all this power and might, he willingly gave his life so that all men will rise from the grave. He redeemed man from sin by buying back and clearing payment for debts that must be paid. He did that which we could not do for ourselves. When he died on the cross, he offered himself as a sacrifice for the laws that would be broken in this mortal life. He bridged the gap between God and fallen man. His pure, without blemish, physical (yet immortal) body was sacrificed to pay the price for man's sins. This great sacrifice is called *the atonement*, which means that the life, suffering, and death of Jesus Christ is what will ultimately reconcile God to man:

> "Yea, remember that there is no other way nor means whereby man can be saved, only through the atoning blood of Jesus Christ." (Book of Mormon, Helaman 5:9)

Although Jesus Christ was a citizen of the world, just as you and I are citizens of the world, his role was of a most awesome nature. He was the Son of an immortal Father, yet born of a mortal mother. Because of this relationship, he did not have to suffer at Gethsemane nor die on the cross of Calvary, for he had within him the power to escape death. Yet Jesus loved mankind with such pure, unqualified love that he never faltered in that which he had been chosen to do. Rather than call forth legions of angels to rescue him from such an agonizing fate, he suffered and died so that mercy could be extended; yet the demands of justice be met. Jesus Christ gave his life as an offering to satisfy broken law: "*Behold he offereth himself a sacrifice for sin, to answer the ends of the law*." (Book of Mormon, 2 Nephi 2:7)

Love, the Motivating Force Behind the Atonement

Heavenly Father allowed His Only Begotten Son to be sacrificed because of His great love for His children. In turn, the Savior unselfishly gave his life because of his great love for his brothers and sisters. He loved with such unqualified love that his every move was for our benefit. Because he knew that only those without sin could return to the presence of the Father, he took upon himself our sins so that we too, might become pure. The giving of his life was the greatest example of love that he could give, for by doing so, he gave his all. Since love is the strongest force in the world, by forfeiting that which was most precious and dear, the Savior was given the power to draw all men unto him:

> "He doeth not anything save it be for the benefit of the world; for he loveth the world, even that he layeth down his own life that he may draw all men unto him." (Book of Mormon, 2 Nephi 26:24)

Adam and Eve: The First to Break the Law

When Eve partook of the forbidden fruit she broke a physical law, therefore a penalty had to be paid. But should she be required to pay the price, even though Adam shortly thereafter partook? Although this act of disobedience brought death and sin into the world, we must never forget she was only doing that which we had all agreed upon in the pre-existence—perhaps not knowing how mortality was to be brought about, but nevertheless agreeing that it was to be so. By eating the forbidden fruit, a physical law was broken; therefore, a penalty had to be paid. Jesus Christ, in essence *said, "I will pay the penalty, I will atone for Adam and Eve's transgression and redeem man from the fall."* By giving his life, he broke the bands that physical death had imposed upon the world, thus assuring that all men would rise from the grave. But who would loose the bands of the *spiritual* death? Would the Savior have the same compassion for those suffering this type of death?

His great atoning sacrifice was two fold; the first part unconditional, the second part on condition of repentance. One might ask why there are again two parts? It is because two types of death had to be considered since two types of bodies must be perfected: the physical and the spiritual. Each is different, therefore, loosening the bands of the spiritual death would be different from loosening the bands of the physical death.

Overcoming Physical Death

Physical death is a deteriorating process that causes physical and spirit matter to no longer bond. It is brought on by weaknesses that we, as well as others, have sown. Things such as hereditary factors, viruses, improper eating, and disease play a part in this separating of body and spirit. But this separation is only temporary, for in time, all will rise from the grave, never again to taste its stinging blows. *This is the unconditional part of the atonement*, for all will be resurrected without effort. It is a free gift from the Father and the Son. This part reveals the Father's mercy and is called salvation, for it provides the way for us to be saved from a death brought on by someone, other than ourselves—Adam and Eve.

> "Now, there is a death which is called a temporal death; and the death of Christ shall loose the bands of this temporal death, that *all* shall be raised from this temporal death." (Book of Mormon, Alma 11:42, emphasis added)

Overcoming Spiritual Death

Spiritual death is different from physical death in that it is brought on by the weaknesses and evil deeds that we ourselves sow within our bodies. It too reflects a deteriorating process, only this time within the spirit part that is bringing us life:

> "And now behold, I say unto you then cometh a death, even a second death,

> which is a spiritual death; then is a time that whosoever dieth in his sins, as to a temporal death, shall also die a spiritual death; yea he shall die as to things pertaining unto righteousness." (Book of Mormon, Alma 12:16)

This is the conditional part of the atonement. The Savior has tried to warn us of the consequences of sin, but should we refuse to heed his warnings and continue to disobey, a price must be paid for our disobedience. Who will pay that price? We are told if we do not repent, it will be us!

> "For behold, I, God, have suffered these things for all, that they might not suffer if they would repent. But if they would not repent they must suffer even as I." (Doctrine and Covenants 19:16-17)

Does this sound like an unloving God? Certainly not, it sounds like a just God who knows there are laws governing his creations and that they must be obeyed, upheld, and sustained if we are to continue existing. If he were unloving, he would not have told us of these laws. Instead, he has declared from the beginning that if we do not repent of our wrongdoing, we will suffer for our own sins. This is because his great atoning sacrifice will not cover man's unrepented sins, for no matter how great his love, he must abide by the laws that govern matter. It is simple; should we persist in wrongdoing, our elements will continue to reflect that disobedience. He will not force us against our will, therefore if we refuse to repent, we will suffer the consequences of our own disobedience! We learn from Helaman 5:10 of the Book of Mormon that he would come not to redeem them *in their sins* but to redeem them *from their sins*.

After the physical death and all have risen from the grave, there will come a time of judgment. Its purpose is to determine the degree to which one will be exalted or damned. It will be a blessing or a cursing, depending upon how well we obeyed. This judgment will reveal the Savior's justice, as it will come as a reward or damnation of our life's work. Since the Savior will atone for repented sins but not unrepented ones, if we have repented, it will reflect that change:

> "And he shall come into the world to redeem his people; and he shall take upon him the transgressions of those who believe on his name; and these are they that shall have eternal life,...*Therefore, the wicked remain as though there had been no redemption made, except it be the loosing of the bands of death*; for behold the day cometh that all shall rise from the dead and stand before God, and be judged according to their works." (Book of Mormon, Alma 11:40-41, emphasis added)

In the Atonement, Both Mercy and Justice are Satisfied

Could the atonement and judgment be opposites that work together as one? It would seem to be so, for the atonement reveals God's mercy, while the judgment reveals His justice. The atonement provides the means whereby man can overcome physical and spiritual death, whereas the judgment sustains and carries out the law. Mercy is in the spiritual realm, while justice is in the physical realm. We find this same order within our elements, for our spirit side wants to give while our physical side wants to be filled and satisfied. The law is again found "in the structure and behavior of the very smallest things," for the atom holds the story of creation and the plan of salvation in microcosm.

There is perfect balance in the Savior's atonement. The law is upheld and sustained; yet mercy is offered to those who desire to repent:

> "And thus we see that all mankind were fallen, and they were in the grasp of justice; yea, the justice of God, which consigneth them forever to be cut off from his presence. And now, *the plan of mercy could not be brought about except an atonement should be made*; therefore God himself atoneth for the sins of the world, to bring about the plan of mercy, to appease the demands of justice, that God might be a perfect, just God, and a merciful God also.... *What, do ye suppose that mercy can rob justice?* I say unto you, Nay; not one whit. If so, God would cease to be God. And thus God bringeth about his great and eternal purposes, which were prepared from the foundation of the world." (Book of Mormon, Alma 42:14,15,25-26, emphasis added)

Our Heavenly Father's gift to man is free agency and a Savior who will atone for the sins of those who desire to be saved. But he cannot save us against our own desires, for that would be compelling us against our own will. To extend the atonement to those who do not want to repent and seek forgiveness would be to extend mercy against one's own agency. Since man has control over his own destiny, he can choose to obey or disobey, but justice requires that only repented sins benefit from the conditional part of the atonement. He, who chooses not to repent, will not receive the blessings that come from repenting.

All things must be upheld and sustained by law; otherwise, if the law is not binding we will end up in chaos. This was the purpose of the Savior's infinite atonement. It was to help man overcome the consequences of broken laws, for a broken law brings change to spirit and physical elements that want to work as one. Because he has taken upon himself our debts, we must repay him in order for the conditional part of the atonement to be satisfied. This repayment requires that we do something to bring about change, for surely if the *atonement* is to have any affect upon our lives, we must bring our elements into an *at-one-ment* with his.

What Must We Do to Enter the Kingdom of God?

First, we must have faith in the Lord Jesus Christ and believe that he was truly sent by our Heavenly Father. Second, we must repent of our sins and cast aside all that is against the nature of God. Third, we must be baptized by immersion for the remission of our sins by one who has the authority. Fourth, we must receive the Gift of the Holy Ghost by the laying on of hands, by one who has the authority. After we have done these things, we must stay faithful to the end. It is true that all will rise from the grave, but only those whose elements are in a state of wholeness, will dwell with the Father in eternal light.

Jesus Christ loves us perfectly in every way. His great atoning sacrifice made it possible for us to overcome the effects of the physical death, brought on by Adam and Eve partaking of the forbidden fruit. It also provided the means whereby we can overcome the effects of the spiritual death, brought on by our individual sins. His atonement is a testimony that he is the Savior of the world and that he works within a framework of law. Had he taken upon himself man's unrepented sins, he would have gone against the justice that must be met. He would have to force our elements back into their original state, thus making us puppets, manipulated at will, instead of the responsible individuals he wants us to become.

WE MUST REPENT, BUT LITTLE CHILDREN NEED NO REPENTANCE

What does this have to do with the atom? If the balance between positive and negative energy is disrupted, it must be regained. If it has not, then there is no need of restoring it.

How important is repentance? We learn from this scripture that:

> *"All men, everywhere, must repent, or they can in nowise inherit the kingdom of God, for no unclean thing can dwell there, or dwell in his presence."* (Pearl of Great Price, Moses 6:57)

This is plain enough. Repentance is important. But what is repentance? The dictionary says it is to feel much sorrow for sin or fault as to be disposed to change one's life for the better, is penitent. It means to change from doing what is wrong to doing what is right. To begin the process, we must first feel sorrow for what we have done; otherwise we will see no need for change. This change of attitude will then bring a desire to change our actions.

Repentance has to do with this mortal state, for when we sin, we do it with our physical bodies. Since excessive negatives are contrary to the nature of righteousness, it is an essential part of God's plan that this course of disobedience be reversed:

> "Nevertheless there was a space granted unto man in which he might repent; therefore this life became a probationary state; a time to prepare to meet God." (Book of Mormon, Alma 12:24)

To return to the presence of our Heavenly Father we must rid ourselves of those things that He abhors: things such as dishonesty, immorality, hypocrisy, pride, and selfishness. Each creates an overabundance of negatives within the body, therefore, if we do not repent of our sins, we will be damned, or in other words our progress stopped. By repenting, we pull out the obstruction that is in our way, thus making it possible to move forward.

Repentance is an ongoing process. By learning to obey a law in one area, we are then given challenges in another. Repentance is a positive because it is an ever-expanding process that leads upward. This would explain this next scripture:

> "Blessed are those who hearken unto my precepts, and lend an ear unto my counsel, for they shall learn wisdom; *for unto him that receiveth I will give more; and from them that shall say, We have enough, from them shall be taken away even that which they have."* (Book of Mormon, 2 Nephi 28:30, emphasis added)

This in essence is saying that obedience moves us in a forward direction that adds positives. But if we do not repent and hearken unto his words, it is like the negative that takes away. Does this sound like we sought for things that would not satisfy by creating excessive negatives that needed to be filled? Always wanting "give me" elements without having the proper balance of "I'll give" and "I'll share?"

Repentance gives hope where once there was despair, confidence where once there was discouragement, faith where once there was unbelief. Could anything give more hope than to know our sins can be forgiven? The Lord has said, "Come now, and let us reason together...*though your sins be as scarlet, they shall be as white as snow; though they be red like crimson, they shall be as wool*." (Bible, Isaiah 1:18)

Little Children Need No Repentance

The Savior qualifies his message of repentance by saying that little children need no repentance. This seems contradictory, for why should they be exempt from doing that which is required of all men? Could it possibly have something to do with their elements?

> "...Wherein the sins of the parents cannot be answered upon the heads of the children, for they are *whole* from the foundation of the world." (Pearl of Great Price, Moses 6:54, emphasis added)

The Lord speaks of children as whole, while sinners as sick and in need of a physician:

> "Listen to the words of Christ, your Redeemer, your Lord and your God. Behold, I came into the world not to call the righteous but sinners to repentance; the whole need no physician, but they that are sick; wherefore, little children are whole, for they are not capable of committing sin." (Book of Mormon, Moroni 8:8)

Why are little children whole and incapable of committing sin? Since whole means complete, entire, or total amount, I believe little children are whole because the strength of their spirit matter is equal to the strength of their physical matter. They need no physician because their two bodies are working together as one. Only those whose elements have lost their ability to work in concert need a physician or healer to bring back their wholeness. If we think of wholeness as a circle or a sphere, we can see how everything pulling together, as in the atoms, would be producing balance and wholeness. Remember the silhouettes of the inert gases? They were balanced and whole because their positive and negative parts were of near equal strength. Would the same be true of little children? Would they too be like an inert gas, unable to be affected by an outside influence, if their spirit and physical bodies were balanced and whole?

Perhaps those who have seen the inert gases in a see through glass tube will better understand this characteristic of wholeness and balance. Each gas is colored a particular color in order to distinguish it from the others, yet there are no barriers separating them. You would think under these conditions they would all blend into one gas, but this is not the case. The one next to them does not influence them and they do not want to act upon any of the others. They neither want to give or share electrons or receive electrons since their atoms are complete and whole. They have no empty spaces that need electrons, nor do they have electrons jutting out that they want to give or share.

Should children's elements be in a state of wholeness, Satan could not influence them, for there would be no need to "act" or be "acted upon." Since Satan is an outside force trying to disrupt their lives, he would have no power over them, for their elements would not respond to his enticements. This would explain this next scripture:

> "But behold, I say unto you, that little children are redeemed from the foundation of the world through mine Only Begotten; *Wherefore, they cannot sin, for power is not given unto Satan to tempt little children*, until they begin to become accountable before me." (Doctrine and Covenants 29:46-47, emphasis added)

When do children become accountable?

> "And their children shall be baptized for the remission of sins when *eight years* old and receive the laying on of hands." (Doctrine and Covenants 68:27, emphasis added)

The courts of the land also recognize the testimony of children when they are eight. They know that by then, they are able to recognize right from wrong, as well as understand the consequences of good and bad behavior. What could have caused this change of innocence? Could it be, as they grow and develop, the energy of their physical bodies starts to exceed the energy of their spirit bodies, thus causing an imbalance between the two? Eating too many acid type foods could be part of the problem, since most children would rather have a hamburger than a dish of vegetables or a tasty fruit. And how can those who are physically abused and degraded not be affected by the terrible injustices they are subjected to? Granted, it is all very complex and there are many factors in the equation, but still, should excess negatives in whatever form start entering their systems, it would upset the balance between two bodies that are working together as one.

We Must Become as a Little Child

Is it possible to reach a state of wholeness wherein we too can become as a little child? The Savior has said:

> "And again I say unto you, ye must repent, and be baptized in my name and become as a little child, or ye can in nowise inherit the Kingdom of God." (Book of Mormon, 3 Nephi 11:38)

How can we, who have taken on the ills of the flesh, become as a little child? The above scripture tells us that not only must we become as a little child, we must also repent and be baptized in Christ's name. The words "repent and be baptized" hold the key to this enigma. One is in the spiritual realm and the other in the physical realm.

Put them together and they bring wholeness to our being. How is this so? When we repent, we stop putting excessive negatives into our bodies and when we are baptized (and honor this covenant by doing what the Savior would do), we create positives. By stopping the excessive flow of negatives and adding an abundant supply of positives, balance is restored to our being.

Since our bodies have already taken on the ills of the flesh, we will never again look like a little child, but by throwing off the ways of the natural man and yielding to the enticements of the Holy Spirit we can become as a little child:

> "For the natural man is an enemy to God, and has been from the fall of Adam, and will be, forever and ever, unless he yields to the enticings of the Holy Spirit, and putteth off the natural man and becometh a saint through the atonement of Christ the Lord, and *becometh as a child*, submissive, meek, humble, patient, full of love, willing to submit to all things which the Lord seeth fit to inflict upon him, even as a child doth submit to his father." (Book of Mormon, Mosiah 3:19, emphasis added)

As a child, we were obedient to our earthly father. To become perfect, we must be obedient to our Heavenly Father. To become as a child, we must learn to submit our will to His will and willingly do whatever He requires. This obedience will bring forth the blessings of heaven and cause us to again become whole. With our two bodies working together as one, we will no longer be affected by Satan's devious influence, for our elements will not allow it.

WE MUST BE BORN AGAIN

What does this have to do with the atom? The atom is composed of both positive and negative energy. When "born again" our negative physical body and positive spiritual body will work again as one, as they did at birth.

In the Bible, when a man by the name of Nicodemus asked Jesus what he must do to be saved in the kingdom of God, Jesus answered, *"Except a man be born again, he cannot see the kingdom of God."* (Bible, John 3:3) Nicodemus was perplexed. He then asked how a man could be born when he is old? And could a man enter the womb a second time to be "born again?" In this next scripture the answer is given on why we must be "born again" to inherit the kingdom of God:

> "And the Lord said unto me: Marvel not that all mankind, yea, men and women, all nations, kindred's, tongues and people, must be born again; yea, born of God, *changed* from their carnal and fallen state, to a state of righteousness, being redeemed of God, becoming his sons and daughters. And...unless they do this, they can in nowise inherit the kingdom of God." (B of M, Mosiah 27: 25,26, emphasis added)

"Born again" means to change from a worldly, carnal state, to a state of righteousness. To be "born again," we must receive by the spirit a powerful awakening that Jesus is the Christ, the Savior and Redeemer of the world, and that only through him can man be saved from the ill effects that sin has imposed upon the world. This truth must touch our spirits with such power that it gives us the strength and desire to change our lives to conform with his life and teachings. If we do so, our actions will reflect righteousness, rather than carnal behavior and our desire will be to do good continually.

> The prophet Alma has said, "Have ye experienced this mighty change in your heart?" (Book of Mormon, Alma 5:14)
>
> The prophet Mosiah has said, "The Spirit of the Lord Omnipotent...has wrought a mighty *change* in us...that we have no more disposition to do evil, but to do good continually." (Book of Mormon, Mosiah 5:2, emphasis added)

This powerful awakening of the spirit is only the beginning of this change, for the Savior then tells Nicodemus:

> "Except a man be born of the water and of the spirit, he cannot enter the kingdom of God" (Bible, John 3:7)

After being spiritually awakened and brought to the realization that only Jesus Christ, the Savior and Redeemer of the world, has the power to redeem us from our fallen state, we must then be born of the water and the spirit, meaning we must be baptized and given the Gift of the Holy Ghost. This again reflects that there is both a physical and a spiritual part to this process of change.

Two Births – The Physical and the Spiritual

Our first birth was a physical birth that left us mainly with concerns of the flesh. But how could it be otherwise—for when entering this physical world, our mothers fed and diapered us and took care of our every need. As a result, things like "what will I eat," "where will I sleep," and "what will I wear" took precedence in our lives. Our natures became like the negative elements that are always wanting to be filled and satisfied.

Our second birth is a spiritual birth—one that leaves us mainly with concerns of the spirit. We then want to give and share, just like our spirit bodies that are made of positive elements, want to give and share. We are born again once our spirit bodies have become the dominant part that restrains our physical bodies from doing evil.

During our first birth, we enter into our individual families. During our second birth, we enter into the family of God. Since there is an opposition in all things, our physical birth is carried out by a mother, our spiritual birth by a father. This makes for a wholeness and oneness, like the beginning set pattern within the atom.

Each birth has three essential ingredients: water, blood, and spirit. During the first birth, our developing physical body lay in our mother's watery womb; then at a given time, our spirits entered the newly formed body to give it life. To bring about this physical birth, our mothers go through pain and suffering, even spilling blood in our behalf. The second or spiritual birth is brought about in a similar manner. This time our elder brother, Jesus Christ, spiritually begets us. Even as a mother in the throes of childbirth, the Savior went through pain and suffering; ultimately shedding his blood in our behalf:

> "And he cometh into the world that he may save all men if they will hearken unto his voice; *for behold, he suffereth the pains of all men*, yea, the pains of every living creature, both men, women, and children, who belong to the family of Adam." (Book of Mormon, 2 Nephi 9:21, emphasis added)

After shedding his blood at Calvary, Jesus was buried in a tomb. Three days later his spirit again entered his body to give it life. To be "born again," we too must be buried in the waters of baptism so that we might come forth as new creatures that have left the old behind. We too, must be given the Gift of the Holy Ghost by the laying on of hands, so that the spirit may then enter our body to give it life. These saving ordinances are in similitude of the life, death, and resurrection of Jesus Christ and are an essential part of being "born again!"

Jesus Christ has spiritually begotten us, therefore, in a sense he is our Father and all who are spiritually "born again" must come through him:

> "And now, because of the covenant which ye have made *ye shall be called the children of Christ, his sons, and his daughters; for behold, this day he hath spiritually begotten you*; for ye say that your hearts are changed through faith on his name; therefore, ye are born of him and have become his sons and his daughters." (Book of Mormon, Mosiah 5:7, emphasis added)

By taking his good name upon us, we must be willing to defend it against the slanderous words and tenuous actions of the unbelieving. Should we be ashamed and stand idly by when righteous actions are maligned and denigrated, we would not be upholding our part of the covenants and ordinances we have entered into with him. This is part of the responsibility of being spiritually "born again."

We are Not Saved Without Effort

To be "born again" we must follow the pattern set by our Lord and Savior, Jesus Christ. We cannot say words and expect miracles to happen. We must repent and be baptized in his name and change from the natural man to the spiritual man. Our spirits must become the governing part, not our physical bodies that want to be filled and satisfied, for Jesus has said:

> "That which is born of the flesh is flesh; and that which is born of the spirit is spirit. Marvel not that I said unto thee, Ye must be born again." (Bible, John 3:37)

It is interesting that the Savior would refer to the second birth as a "rebirth" or "born again" condition, for how else could we enter the presence of our Heavenly Father unless our two bodies were balanced and whole, as in the beginning?

THE ROOT AND THE BRANCH — THE GIVER AND THE RECEIVER

What does this have to do with the atom? Both the positive and negative are needed for energy, magnetism, and light to be created. When the negative overwhelms the positive, problems emerge.

In Jacob, chapter five of the Book of Mormon, we read of yet another imbalance within the elements. This time the Lord is teaching, by parable, of olive trees that have a pattern of flourishing and then dying in his vineyard. Although it is called the allegory of the wild and tame olive trees, it is in reality speaking of the house of Israel. In the pre-existence, this family of nations was given the responsibility of carrying out the Lord's work here on the earth.

In the beginning, the master of the vineyard (the Savior), carefully planted and cared for the trees, (the house of Israel) so they would bring forth good fruit and accomplish his purposes. The first harvest was successful. However as time went on, the trees became unbalanced. They were lofty and full at the top and began to show signs of dying. To correct the imbalance, the master had the servant prune the branches and dig around and nourish the roots of the trees. In this way, the energy going into the top foliage was reduced and the roots were given additional strength. This same thing was repeated many times in the allegory, and with always the same results. For a while the trees would bear good fruit, then after a time, would begin to die.

At one point, the Lord of the vineyard decided to cut off a good branch and graft it to another tree in a far off part of his vineyard. He was alluding to the tribe of Joseph, whose descendants were transplanted far from the main body of the house of Israel. This branch did well for a time, and then it too, began to die.

The master gave the trees all the tender loving care he could, yet the problem persisted. Toward the end of the allegory, the master told the servant that it would be the last time he would try to save the trees, or house of Israel (emphasis added):

> "But what could I have done more in my vineyard? Have I slackened my hand, that I have not nourished it? Nay I have nourished it, and I have digged about it, and I have stretched forth mine hand almost all the day long, and the end draweth nigh. And it grieveth me that I should hew down all the trees of my vineyard, and cast them into the fire that they should be burned. Who is it that has corrupted my vineyard? And it came to pass that the servant said unto his master: *Is it not the loftiness of thy vineyard—have not the branches thereof overcome the roots which are good? And because the branches have overcome the roots thereof, behold they grew faster than the strength of the roots, taking strength unto themselves.* Behold I say, is not this the cause that the trees of thy vineyard have become corrupted?...Wherefore, let us go to and labor with our might this last time, for behold the end draweth nigh, and this is for the last time that I shall prune my vineyard." (Book of Mormon, Jacob 5:47-48,62)

Since trees are made of the same basic elements as man, we see this same principle of balance and imbalance within the elements applying to all of God's creations. The trees, or house of Israel, were dying because too much energy was going into the top foliage, thus taking from the roots, which were giving them life.

Whether it is the house of Israel, individuals or trees, it is the same. The lofty foliage represents the physical negative part while the root represents the spiritual positive part. Because the root is the giver of life and the branches the receiver of life, when too much left, negative (–) energy is taken in, the balance between the positive and negative is disturbed and undesirable consequences begin to emerge. When balanced and whole, good fruit is produced; when out of balance, bad fruit is produced. The trees (or house of Israel) began to die because their energy was going to the branches, leaving little nourishment for the root.

We are like that. The spiritual part that brings life to our whole body is diminished if too much attention is directed toward the outward, physical things of this world. We bear bad fruit when our spirits begin to be unteachable and full of false pride. This unbalanced condition can be corrected if we eliminate and clear away the bad things that we have allowed to enter our lives:

> "Wherefore ye shall clear away the bad according as the good shall grow, *that the root and the top may be equal in strength*, until the good shall overcome the bad." (Book of Mormon, Jacob 5:66, emphasis added)

Becoming Carnal, Sensual, and Devilish

What are the ways of the world? They are the temptations Satan puts before us in diverse and countless ways. They are excessive negatives that distract us from the spiritual realm. They are perversions (pornography) and unnatural lifestyles. They are gratification of the flesh and fleeting thrills to make the body feel "sensational." They are seeking money for the sake of one's own pleasure. They are prideful power, immodest dress, lying, and taking the sacred name of deity in vain. They are fame brought on by actions that are excessive and crude. They are feeling the need to follow the same crude ways as the famous. These are just a few of the ways of the world and of course are man's ways not God's ways.

These next two scriptures refer to this condition of disobedience as carnal, sensual and devilish:

> "And they loved Satan more than God. And men began from that time forth to be *carnal, sensual and devilish*." (Pearl of Great Price, Moses 5:13, emphasis added)
>
> "Behold Satan hath come among the children of men, and tempteth them to worship him; and men have become *carnal, sensual, and devilish*, and are shut out from the presence of God." (Pearl of Great Price, Moses 6:49, emphasis added)

It is interesting that carnal, sensual, and devilish are used in the same sequence. When studying the definitions, the reason is clear. Each progresses from a lesser state of disobedience to a greater state of wickedness.

Carnal: not spiritual, merely human and temporal, worldly, pertaining to the flesh or the body, its passions and appetites.
Sensual: excessively inclined to the gratification of the senses. Lewd or unchaste means pertaining to or given to the gratification of the senses or the indulgence of appetite.
Devilish: like or befitting a devil, diabolical, fiendish, excessive, very great.

We cannot be devilish before being sensual and carnal because binding our elements is a step-by-step process that takes time. It is the building of one negative (–) upon another negative (–). The first step in this downward spiral is failing to recognize that it is the Spirit that is bringing us life and that it comes from God. This is what the definition of *carnal* implies, "not spiritual, merely human." Some may think we consist of only a physical body that can be seen. However, it is the unseen spirit that receives the restraining impressions from on high. When we fail to receive these whisperings, we are spiritually weakened and are of a carnal, worldly mindset. It is the natural man that does not obey the impressions that come to his spirit:

> "The natural man receiveth not the things of the Spirit of God: for they are foolishness unto him: neither can he know them because they are spiritually discerned." (Bible, Corinthians 2:14)

We become *sensual* when physical concerns become the primary focus in our lives and we become overly concerned with the gratification of the flesh. When emptiness and yearnings arise we center our lives even more on making the body feel good. But the more we try to satisfy these longings, the more excessive our actions become. As one thrill loses its excitement, we move on to greater ones.

We become *devilish* when we no longer have the power to resist evil and when extreme and devious thoughts and actions are a daily occurrence. We are then under Satan's power and our elements bound by sin. We think evil continually because there are now no restraints. We regress from one sin to yet another because our spirit bodies are unable to restrain our physical actions. We may want to change, but having drawn off so much spiritual strength, we no longer have the power to accomplish such a task. This downward spiral will continue until our desire to change becomes greater than our desire to sin.

Bound By Sin

Sin at first seems such a small thing and so innocent that its colors do not really show. But once caught in its vicious tentacles, it has the ability to bind us in an ever-tightening hold. Once restraints are loosened and sin no longer has a stranger's face, we move from one stage to yet another, ever hoping to find the happiness we so desire. But sin never was happiness, it is like the negative element that is not satisfied:

> "For they that are after the flesh do mind the things of the flesh; but they that are after the Spirit, the things of the spirit. For to be carnally minded is death; but to be

> spiritually minded is life and peace. For if ye live after the flesh, ye shall die; but if ye through the Spirit do mortify the deeds of the flesh, ye shall live." (Romans 8:5-6,13)

The Savior has at many times and in many places patiently worked with the house of Israel to produce "good fruit." But it is always the same. When he plants his message in the hearts of men, they believe. Then Satan comes among them, saying, "believe it not" and the scriptures tell us "they believed it not." It is a vicious cycle. Satan's onslaught weakens society. When hopes are dashed that the Savior's message could be true, many no longer hesitate to take on the ways of the world. With their spiritual faith crushed, sin no longer has a stranger's face.

Too Much of the World

The house of Israel was scattered and dispersed because they had become wicked and rebellious. Prosperity was their downfall and has continued to be their challenge throughout the centuries. When obedient and submissive they received great blessings. However, in time, their prosperity turned to their condemnation. They invariably started taking in too much of the physical world and their blessing eventually became a curse. This has been the pattern since the dawn of history. War, famine, and sickness have a way of humbling the soul, but when blessed with prosperity, man no longer feels the need of God's sustaining hand.

And so it is today. Too much of our energies are being spent on physical gratification, thereby taking strength from the spirit part that is giving us life. But remember; even though it is a daily challenge to maintain our spiritual strength when the physical world is so aggressively crowding in upon us, the spiritual is a positive that moves things forward, whereas the physical is a negative that regresses. Unless they are balanced and whole, the physical world works to our condemnation.

Will the Lord be successful in pruning and nourishing this one last time? Will the house of Israel again become one?

> "And they did obey the commandments of the Lord of the vineyard in all things...and the wild branches began to be plucked off and to be cast away; and they did keep the root and the top thereof equal, according to the strength thereof....*and they became like unto one body; and the fruits were equal*...And it came to pass that when the Lord of the vineyard saw that his fruit was good, and that his vineyard was no more corrupt, he called up his servants, and said unto them: Behold, for this last time have we nourished my vineyard; and thou beholdest that I have done according to my will; *and I have preserved the natural fruit, that it is good, even as it was in the beginning*...and have brought unto me again the natural fruit, that my vineyard is no more corrupted, and the bad is cast away." (B of M, Jacob 5:72-75, emphasis added)

In the end, the vineyard will no more be corrupted and the ancient house of Israel will become like unto one body, as it was in the beginning. They will turn from the degrading ways of the evil one and again bear good fruit. They will again become one, as required of the *law of opposites*.

FAITH WITHOUT WORKS IS DEAD

What does this have to do with the atom? Faith is a positive, works is a negative. Together they produce energy, magnetism, and light; alone they have no power.

Do you sometimes wonder what place "works" has in the eyes of the Lord, for many say there is nothing required of us but to confess Christ and his grace is sufficient to save. Even the scriptures present two different points of view:

> "For by grace ye are saved through faith; and that not of yourselves: it is the gift of God: Not of works, lest any man should boast?" (Bible, Ephesians 2:8-9)

But can we disregard its opposite?

> "Wherefore, my beloved, as ye have always obeyed...work out your own salvation with fear and trembling." (Bible, Philippians 2:12)

Even though these two scriptures seem to contradict one another, they do not, for each is a complementary opposite in two different realms; one in the spiritual, the other in the physical.

How does grace fit in with faith and works? We know that grace is the free, unmerited love and favor of God and that without it we could do nothing. It was by His grace that He created the world and permitted us to come here. By His power and light, He organized the elements, thus giving us the air we breathe and the ground we walk on, the sun that shines and the plants that grow. By His grace, He allowed His Son to atone for our sins. He inspires and guides us along life's stormy paths and without Him we are nothing.

Then why do we read that work is involved?

> "For we labor diligently to write, to persuade our children, and also our brethren, to believe in Christ, and to be reconciled to God; for we know *it is by grace that we are saved, after all we can do*." (Book of Mormon, 2 Nephi 25:23, emphasis added)

What is meant by "after all we can do?" We know it is by the atonement of Jesus Christ that we are saved and without this great gift we stand helpless in delivering ourselves from the consequences of sin and death. But there are conditions to this gift. He will deliver us from physical death, but only those who repent and do something to change their lives, will he deliver from spiritual death. It is somewhat like being in the bottom of a deep hole with no way out. We pray for someone to relieve our plight. The Savior shortly thereafter comes along and lowers a ladder into the hole so that we might escape our dilemma. At this point we must do something ourselves. We must climb up the ladder and out of the hole; otherwise his gift is of no value. Faith, works, and grace are like this. We must have all three to be saved from the consequences of sin. But after climbing out of the hole, should we say, "Oh, how great I am, for doing such a marvelous thing" or should we say, "Oh, how great thou art, for providing the means whereby my life could be saved?" Of course the answer is obvious. It is to the Savior that we give praise, for through his grace it is possible.

Our Spiritual and Physical Bodies Must Work Together as One

It seems the relationship between faith, works, and grace is the same relationship we see within the individual atom. Faith has to do with the spirit body that is embodied within the positive energy field that wants to act, while works has to do with the physical body that is embodied within the negative energy field that wants to be acted upon. This next scripture speaks of this relationship:

> "What doth it profit, my brethren, though a man say he hath faith, and have not works? can faith save him? If a brother or sister be naked, and destitute of daily food, And one of you say unto them, Depart in peace, be ye warmed and filled; notwithstanding ye give them not those things which are needful to the body; what doth it profit? *Even so faith, if it hath not works, is dead, being alone... For as the body without the spirit is dead, so faith without works is dead also."* (Bible, James 2:14-17,26, emphasis added)

Some people present scripture that make it sound as if faith is all that is necessary for a man to be saved. They don't bother to mention there are just as many scriptures telling us that works are necessary to be saved. The Apostles wrote the epistles in the New Testament to members in far off places who needed help with a particular problem. It is obvious that the Epistle of James was written to emphasize "works" just as the Epistle to the Romans was written to emphasize "faith" and "grace." Each is important to the principle of salvation for one cannot be put aside as insignificant, while the other is given greatest emphasis.

Perhaps the controversy between faith and works comes from the Bible speaking of salvation in two different ways. But once we realize there are spiritual and physical parts to consider, the confusion ends. Salvation, when used in the context of being resurrected from the grave, is a free gift that comes to all. However, when used in the context of being exalted, requires work. Otherwise what is the purpose of works, if all are saved without effort? To be exalted means to rise above what we are now. It means living the laws of a higher state of existence so that our elements will be inseparably connected with the glory of the Son. This moving forward to greater opportunities requires the kind of faith that leads to works; otherwise there is no significant change to elements that should be working as one. Faith alone does not have the power to exalt man. It cannot be accomplished by simply saying, "I believe," for it takes effort on the part of every man seeking this glorious reward.

But what must we have faith in? The answer is Jesus Christ, for he is the author and finisher of our faith and has the ability to save us from death. He is the vine from whence we, the branches, receive strength. He must be our foundation; otherwise we cannot stand. And how can faith in him do this? With the spiritual force within us firmly rooted in Christ, we can move forward in a positive direction, activating physical elements into serving others and doing "good works." To believe that Jesus Christ will save us with no thought or action taken to obey his words is not following natural law. Faith is a forward moving spiritual positive. As it gives and shares its electrons, it activates physical elements that need to be "acted upon." We cannot say we believe and yet do nothing, for faith is a principle that requires its possessor to take mental or physical action:

> "Now faith is the substance of things hoped for, the evidence of things not seen...But without faith it is impossible to please him: for he that cometh to God must believe that he is, and that he is a rewarder of them that diligently seek him." (Bible, Hebrews 11:1,6)

Can you see why the Lord would say?

> "Not everyone that saith unto me, Lord, Lord, shall enter into the kingdom of heaven; but he that doeth the will of my Father which is in heaven?" (Bible, Matthew 7:21)

It is only those who "doeth" that keep their spiritual and physical energies activated. By combining faith and works the whole body is energized.

We are Saved By Grace, After All We Can Do

How do we become perfected? We move from grace to grace until the perfect day. This means that when the Savior puts one truth before our eyes, it will be the foundation for yet another. Depending on our obedience, the full picture will unfold before our eyes. This next scripture speaks of this grace:

> "For if you keep my commandments you shall receive of his fullness, and be glorified in me as I am in the Father; therefore, I say unto you, *you shall receive grace for grace*." (Doctrine and Covenants 93:20, emphasis added)

As long as we labor and continue doing something about our promptings, our Heavenly Father will continue to instruct us. But if we refuse to listen, these blessings will stop. Grace is God's unconditional gift to man, but His gift is of no value unless we do our part. We are told:

> "But by the grace of God I am what I am: and his grace which was bestowed upon me was not in vain; *but I labored more abundantly than they all: which not I, but the grace of God which was with me*." (Bible, 1 Corinthians 15:10, emphasis added)

We all have weaknesses, but only through His grace will we be able to overcome them:

> "Nevertheless, the Lord God showeth us our weakness that we may know that it is by his grace, and his great condescension's unto the children of men, that we have power to do these things." (Book of Mormon, Jacob 4:7)

It is important to know that as we face the trials that life has to offer the Lord will buoy us up in our efforts to succeed. He will not leave us alone, but will aid and strengthen us in our need. His grace will carry us through this mortal existence. By the grace of the Father and the Son, they will help us reach perfection. Not all at once, but here a little, there a little, until we have overcome all things:

> *"Behold ye are little children and ye cannot bear all things now; ye must grow in grace and in the knowledge of the truth."* (Doctrine and Covenants 50:40)

When faith, works and grace are intertwined into one strong, unbreakable cord, our elements are refined and our spirit's made pure. One alone will not do it, as it takes the strength and power of all three.

ETERNAL PROGRESSION IS LIKE THE POSITIVE MOVING FORWARD

What does this have to do with the atom? Eternal progression is forward moving like the positive field of the atom.

Eternal progression is a forward moving sequence of incremental developments. Like the infinity pattern within the atom, it is without beginning or end. Because the positive force is stronger than the negative force, this normal state of the elements gives evidence that progression is a part of God's eternal plan. It is the way the elements have been organized. But we as human beings are unique. We have been given the freedom to progress or regress in this mortal state. We can take the path that enlarges our potential (+) or the one that diminishes and takes it away (–).

To progress, we must first know how to proceed. An exemplar was needed to show us the way. This is why Heavenly Father sent Jesus Christ, His Only Begotten Son, to earth. He knew this valiant son would be faithful to His will and not fall to Satan's temptations. This is why He called him, "The Way, The Truth, and The Life." Time has shown that he was the perfect example. Although born two thousand years ago in an obscure village called Bethlehem, his life, death and resurrection were the most important events to occur in the history of the world. Even time is measured by his life: B.C. meaning Before Christ, and A.D. meaning Anno Domini, Latin for in the year of our Lord.

Christ's perfection does not mean that he stood apart from the world as an ascetic, without the normal experiences of life, for he "*...was in all points tempted like as we are, yet without sin*" (Hebrews 4:15). But the miracle of it all is that even though he was perfect and always chose the right; it was he who said that we could become like him!

> *"I would that ye should be perfect even as I, or your Father who is in heaven is perfect."* (Book of Mormon, 3 Nephi 12:48).

We must therefore look to him for guidance and direction, for through his own experiences he was able to show how perfection might be reached.

Jesus Received Not a Fullness at First

But even Jesus did not receive a fullness at first:

> "And he received not the fullness at first, but continued from grace to grace, until he received a fullness; *And thus he was called the Son of God, because he received not of the fullness at the first.*" (Doctrine and Covenants 93:13-14 , emphasis added)

He received a fullness at the time of his baptism, when the Holy Ghost descended upon him in the form of a dove:

> "And I John, bear record, and lo, the heavens were opened, and the Holy Ghost descended upon him in the form of a dove,...and there came a voice out of heaven saying: "This is my beloved Son." *And I, John, bear record that he received a fullness of the glory of the Father*; And he received all power, both in heaven and on earth, and the glory of the Father was with him; for he dwelt in him." (Doctrine and Covenants 93:15-17, emphasis added)

We are told in this same scripture that if we, too, are faithful, we shall receive a fullness like this faithful son:

> "And it shall come to pass, that if you are faithful you shall receive the fullness of the record of John. I give unto you these sayings that you may understand and know how to worship, and know what you worship, *that you may come unto the Father in my name, and in due time receive of his fullness*. For if you keep my commandments you shall receive of his fullness, and be glorified in me as I am in the father; therefore, I say unto you, you shall receive grace for grace." (Doctrine and Covenants 93:18-20, emphasis added)

Fullness means to be whole, complete and full. (Think of the atom that is complete and whole.) When the scriptures say Jesus received fullness, they are saying that as a son, he has now possessed the complete glory of the Father. How amazing! This means that if we too are faithful, we will in due time receive the same fullness and enjoy the same father-son or father-daughter relationship.

Jesus Could Not Have Been the Messiah Before He Was First the Son

As evidence that there was a progression in the Savior's life that we sometimes fail to recognize, look at the following names by which he is known: Lamb of God, Redeemer, Advocate, Firstborn, Jehovah, Good Shepherd, Elder Brother, Judge, Creator, Only Begotten, Light of the World, Counselor, King of Kings, Anointed One, Messiah, Deliverer, Son, Mediator, and Savior.

Now rearrange them in chronological order and you will see him progressing from one state to another:

FIRST BORN — SON — ELDER BROTHER — ANOINTED ONE — CREATOR — JEHOVAH — ONLY BEGOTTEN — LIGHT OF THE WORLD — THE MESSIAH — MESSENGER — COUNSELOR — GOOD SHEPHERD — LAMB OF GOD — REDEEMER — SAVIOR — MEDIATOR — DELIVERER — ADVOCATE — JUDGE — THE GREAT HIGH PRIEST — KING OF KINGS.

Jesus could not have been the Messiah before he was first the Son and he could not be our Judge before he was first our Savior. Through a day-by-day process of obedience, he progressed from one responsibility to an even greater one. Eternal progression does not mean Jesus will go on to amass new knowledge or to learn new laws, for he already knows all things, since he is the one who created the world in the first place. His progression and glory comes from his creations moving forward. When we put effort into moving his work forward, we too move forward; this is the nature of eternal progression. We have not been given the same responsibilities as Jesus was given. Yet this does not mean we have not been given some responsibility or that we will not be given more, for Heavenly Father knows our potential to accomplish great things.

Suffering: The Refiners Fire

The Savior's perfection did not come without price:

> "Though he were a Son, yet learned he obedience by the things which he suffered; And being made perfect, he became the author of eternal salvation unto all them that obey him." (Bible, Hebrews 5:8-9)

Just as the Savior learned obedience by the things he suffered, to become the sons and daughters of God, we, too, must learn obedience through the things we suffer. It is part of our eternal progression. This next scripture speaks of this suffering:

> "My Son, peace be unto thy soul; Thine adversity and thine afflictions shall be but a small moment. And then, if thou endure it well, God shall exalt thee on high." (Doctrine and Covenants 121:7-8)

Here in mortality we are being refined and perfected so we may live with heavenly parents who are refined and perfected. We should not become discouraged or curse God for our trials, because to live with such exalted beings, our elements must be made strong in order to withstand their glory. They, the elements, must be refined and made pure, free from impurities that are course and debasing. This refining process is much like the refining of precious metals—silver and gold. The metals go through an intense heating and boiling process so that their worthless, undesirable materials will rise to the surface, there to be skimmed off and thrown away. Sometimes fuller's soap is added to pull out the course, impure material and to thicken and strengthen. This process is repeated until the desired purity is reached. The Lord speaks of our being refined in a similar manner:

> *"For, behold, I have refined thee, I have chosen thee in the furnace of affliction."* (Book of Mormon, 1 Nephi 20:10)
>
> "Who may abide the day of his coming, and who shall stand when he appeareth? For he is like a refiner's fire, and like fuller's soap. And he shall sit as a refiner and purifier of silver and purge them as gold and silver." (B of M, 3 Nephi 24:2-4)

Jesus Christ's life was not one of comfort and ease, yet he showed how all men could become perfect through trials that refine and sufferings that make pure. Since he is our exemplar, we must pattern our lives after his. Knowing he was the only man without sin should not discourage us in our quest for perfection, for Heavenly Father knows that we, too, have potential. He knows we will make mistakes and perhaps "kick against the pricks," but that through repentance and long suffering, we can become like His faithful Son.

Eternal Progression Comes Through Jesus Christ

It is through Jesus Christ that we progress, for had he not atoned for our sins, we would have remained forever under the curse of a broken law:

> "Yea, behold I say unto you...there is none other name given under heaven save it be this Jesus Christ...whereby man may be saved." (Book of Mormon, 2 Nephi 25:20)

It was his great atoning sacrifice that saved us from our fallen state. Heavenly Father chose him in the pre-existence to be His emissary and representative here on earth because He knew he would represent Him, as though He were here. Everything pertaining to this world has been done through him. He is the personification of perfection and the only sure way. At the judgment time, every knee shall bow and every tongue will confess that Jesus Christ, is the Lord of the whole earth. All flesh will then acknowledge him and submit to his will.

THE SON OF GOD

What does this have to do with the atom? The Savior is like the positive element that wants to give and share; therefore we know Heavenly Father has this same characteristic.

Satan has always tried to confuse the world as to the true relationship between the Father and the Son. At one point, even Philip, one of the Savior's chosen apostles, did not understand this relationship:

> "Philip saith unto him, Lord, shew us the Father, and it sufficeth us. Jesus saith unto him, Have I been so long time with you, and yet hast thou not known me, Philip? *he that has seen me hath seen the Father*." (Bible, John 14:8-9, emphasis added)

Why is it important to know whom we worship? By knowing their attributes and characteristics, we can then emulate their ways:

> "And this is life eternal, *that they might know thee* the only true God, and Jesus Christ, whom thou hast sent." (Bible, John 17:3, emphasis added)

To know God the Father and His Son Jesus Christ we must know what they are like. Even though the scriptures declare they are two separate and distinct beings, many continue to believe that the two are really one and the same being. But can you visualize in your mind's eye, two distinct personages with the ability to reason, to think and to do, living together within the same confining structure? It is inconceivable. We, of course, have a physical body and a spirit body that co-exist, but only the spirit body which holds the intelligence, reasons, and thinks when the two are separated. The physical body does not have life within itself.

Some say God is an essence of spirit or is unfathomable, yet this next scripture declares we can become like Him: "*Be ye therefore perfect, even as your Father which is in Heaven is perfect.*" (Bible, Matthew 5:48)

How can we become like someone we cannot comprehend; someone without body, parts or passions, or a multiple being in one? If in our minds He is just an all-pervasive force, we could not entertain such a possibility. But if we knew He had a physical body with parts and passions that were real, we could envision such a promise being true.

The concept of the Godhead being incomprehensible seems to have been one of the leading causes of the early day apostasy. There was so much contention concerning the nature of the Father and the Son that the Emperor Constantine called the Bishops together in 325 A.D.. to resolve these differences. The Nicene Creed, which is the foundation of Christianity today, was the result of this gathering. It is called the doctrine of the Trinity, meaning there are three Gods in one, yet three separate Gods.

Perhaps this next scripture is what led many to believe that Heavenly Father and Jesus Christ were one and the same being:

> "Holy Father, keep through thine own name those whom thou hast given me that they may be one, as we are... That they all may be one; as thou, Father, art in me, and I in thee, *that they also may be one in us*." (Bible, John 17:11,21, emphasis added)

This oneness was not of body, but of two personages who see eye to eye on all things. They work in unison and harmony. They have oneness of purpose. They do not contend, but agree on all things. To further show that the Savior was not speaking of the two becoming one in substance, the scripture also says that "we too may become one with them." Should we look on this oneness as one in body or element, we would then have to say our elements must become meshed with their elements. Now this would be inconceivable. This oneness means that all things are working together in harmony as seen within the atom. With the elements working together as one, shouldn't everything else be following suit and doing the same?

The True Nature of the Godhead

From this next scripture we learn of the physical nature of the Godhead:

> "The Father has a body of flesh and bones as tangible as man's; the Son also; but the Holy Ghost has not a body of flesh and, but is a personage of spirit. Were it not so, the Holy Ghost could not dwell in us." (Doctrine and Covenants 130:22)

The Holy Ghost is also a member of the Godhead, but remains in spirit form so that he can testify to our spirits of the Father and the Son. Joseph Smith's first vision gives additional information on the Father and the Son. Joseph describes Heavenly Father as a glorified individual and that we have been made in His image and that Jesus Christ was His Only Begotten Son, separate and distinct from the Father. Their relationship is simple, yet the world has made it hard to understand.

To further clarify the nature of God, I would ask, "Is He loving and kind or does He demand retribution when His laws are broken?" Since we now understand the importance of opposites, it is reasonable to assume that He also has two natures: one that is merciful and one that demands justice. Throughout the scriptures we see these two natures intertwined with the other. Since God and Heavenly Father are one and the same Being, when He tells of the consequences that come when His laws are broken, His words resound like thunder and our thoughts are subdued. But the name Heavenly Father invokes a different response within us. It brings this Great Being down to a more personal, intimate level. He is then our Father, confidant and friend. He is all caring, merciful, and kind.

Now some might think His two natures are at odds with one another, but they are not, for with one, justice is served, while with the other, mercy is extended. We may want to discount the consequences that come from negative behavior, but breaking God's laws are part of the equation. Can you see why it is important to know whom we worship? Once we understand their physical attributes and natures, we can then seriously consider what they are saying.

The Savior Was Sent By the One Great God

Heavenly Father had His Only Begotten Son represent Him here on earth, just as you would send one of your own children to represent you in some far off land. When speaking to his Heavenly Father, Jesus Christ testifies of this reality:

> *"I have glorified thee on the earth: I have finished the work which thou gavest me to do*...Now they have known that all things whatsoever thou hast given me are of thee. For I have given unto them the words which thou gavest me; and they have received them, and have known surely, that I came out from thee." (Bible, John 17:4,7-8. emphasis added)

While on earth, the Savior tried to teach the people that he was not the Father, but was sent by the Father and was representing Him here on earth: *"My doctrine is not mine, but his that sent me"* (Bible, John 7:16) Jesus Christ was God's emissary here on earth: his spokesman. What message did He want us to hear? That His Beloved Son atoned for the sins of the world so that mankind might again enter His presence:

> "Behold I have given unto you my gospel, and this is the gospel which I have given unto you—*that I came into the world to do the will of my Father, because my Father sent me*. And my Father sent me that I might be lifted up upon the cross; and after that I had been lifted up upon the cross, that I might draw all men unto me, that as I have been lifted up by men even so should men be lifted up by the Father, to stand before me, to be judged of their works, whether they be good or whether they be evil— ...And no unclean thing can enter into his kingdom; therefore nothing entereth into his rest save it be those who have washed their garments in my blood, because of their faith, and the repentance of all their sins, and their faithfulness unto the end. Now this is the commandment: Repent, all ye ends of the earth, and come unto me and be baptized in my name, that ye may be sanctified by the reception of the Holy Ghost, that ye may stand spotless before me at the last day. Verily, verily, I say unto you, this is my gospel...; *for the works which ye have seen me do that shall ye also do*." (Book of Mormon, 3 Nephi 27:13-14, 19-21, emphasis added)

We, Too, Can Become a Son or Daughter of God

Heavenly Father is the father of our spirits. This is why we call Him Heavenly Father. Yet the scriptures declare we can become His sons and daughters. This implies there is a process we must go through before we actually become such. Since we already have a spiritual connection, it is evident there must also be a physical connection. We must now show through our physical actions, that we have become like Him.

Since Jesus Christ was the only one physically born of the Father, he was the one chosen to physically show us the way. His life and his words bear witness that he has done this in a most remarkable way. He has shown us how love, patience, and kindness win out in the end and how purity and self-mastery bring peace to the soul. By fearlessly defending the Father's work, he has taught us that we must never fear the rejection of the world. He said that all men must repent and that empathy and forgiveness are part of that path. He has shown us that without charity, we are nothing. His life was built on a foundation of truth and compassion was the fruit of his heart. Can we now comprehend what it means to become like our Heavenly Father? It is really quite simple. When we do what the Savior would do, we become like him, and when we become like him, we become like our Heavenly Father. What a glorious, comprehensible truth!

— SECTION FOUR—ONENESS COMES FROM OBEDIENCE TO LAW —

Will It Be Balance in All things—or—Koyaanisqntsi?

What does this have to do with the atom? It shows that God intends all things to be balanced and whole, just like the elements from which we are made.

The human body stands as a witness that God intends all living things to be balanced and whole, just like the elements from which we are made. We have two ears to hear, two eyes to see, and two nostrils to smell. To keep this balance, we have two arms and two legs, ten fingers and ten toes: five on each hand, five on each foot. In addition, we have three body openings, one for feeding it nourishment, the other two for eliminating its waste. We have two life centers: the heart and the brain. The heart provides mercy, the brain, justice. Truly man is an amazing creature! He has been formed by an all-knowing Father who knows that opposites bring balance and harmony to the whole being.

What is happening in the world today? Are man's ways disrupting this balance of energy? Has "Koyaanisqntsi," the Indian expression for man out of balance, come to pass? Are God's laws being carelessly brushed aside as something of little consequence?

Could Music Be Upsetting This Balance?

It can hardly be disputed that society made a radical shift in its behavior at the same time new sounds began falling on our ears. High standards began to fall when music started to change. Elvis Presley with his gyrating body movements and the Beatles with their new sound sparked a movement in the 60s that took music (depending on one's point of view) from the "good, to the bad, to the ugly." Shortly thereafter, hard or acid rock music appeared on the scene. With this change came the drug culture, commonly known as "the hippie movement." Moral laxness and immorality became a major part of the growing up scene and areas formerly taboo became acceptable at an alarming rate.

Could this sudden upheaval in behavior be a consequence of laws within our bodies being broken? Could the law of opposites be governing music, just as it is governing the Pawlowski Atom? Let us consider: with the treble going up the keyboard (+) and the bass going down (–), it too has a positive part that builds and a negative part that takes away. It too, has a modulating down, like the transitionals stepping down into the inner shells and it too looses its balance and harmony if there are too many bass notes (–) being played.

Discordant, acid, or hardrock music with their notes in disarray and their heavy beats coming from the left side of the keyboard could be called agitating and excessive. Their vibrations are so strong that one can actually feel them pulsating throughout the body. One cannot hum their melodies, for the sounds are like shattered atoms that no longer have the ability to work as one.

Classical music is the opposite. It produces feelings of peace and well being. Like the positive element that soars into infinity, it carries our souls to great heights. Its notes are soft and loud, fast and slow, and treble and bass. With its timing and sounds working together as one, it has balance and harmony.

Evidence of a relationship between music and the elements was found in an article in the *Arizona Republic* (April 5, 1986). It supported the premise that; "In the structure and behavior of the very smallest things, lies the explanation of all matter." It states that when the DNA structure was unraveled over a quarter a century ago, no one would have thought the genetic blueprint for life would also serve as a source of music. (DNA determines what we will look like, our susceptibility to certain diseases and our life span).

Yet this was the finding of Dr. Susumo Ohno, a distinguished scientist at the City of Hope Medical Center. He claimed to have discovered a link between music and the chemistry of genes. Dr. Ohno called his discovery the "primordial secret." He had no formal music training, but knew the basic position of the notes and nothing more. By taking the DNA of cancer genes, genes for antibodies, and those that code for enzymes, and translating them into musical notes, he came up with melodies that are as different as night and day. Some melodies were light hearted and could be likened to the Baroque and Romantic periods while others were like the austere dissonances of composer Igor Stravinsky. He said:

> When played, a cancer-causing gene called an oncogene sounds somber, slightly out of control. An enzyme that breaks down lactose, simple milk sugar, sounds like a lullaby. The melody of a synthetic antibody responsible for detoxifying the chemical phenol sounds lively, lighthearted. Chopin's *Funeral March* resembles a cancer-causing gene.

Dr. Ohno founded his work on the hypothesize that all nature is characterized by periodicities and repetitions—

> Earth spinning on its axis, the revolutions of the planets around the sun. There are nights and days and there are four seasons in the year while the moon waxes and wanes. Even mating calls of songbirds have distinct periodicities in the form of repeated melodies.

Dr. Ohno says the process of correlating the basic chemistry of genes to notes is very simple:

> The relationship is found between the molecular weights of the purine-pyrimidine bases... and the position of notes on the staff...heavier molecular weight mole-

cules assume the lower positions and lighter-weight molecules are assigned higher positions.

He then asks, "Why do we like certain melodies?"

A Culture Without Restraints Creates Bondage

Science has learned how to fragment the atom, but in turn, Satan has taught us how to fragment the elements from which we are made! Should anyone doubt that music affects our elements, why do those who embrace acid or hard rock music, often embrace such outlandish opposites as yellow hair, purple hair, spiked hair, and shaven hair—oversized clothes, undersized clothes, battle fatigues, and bikinis? Does their loud, extreme, and excessive behavior not give evidence that they have embraced a lifestyle that creates a chain of negatives?

Should one take exception with this statement, what kind of behavior do we find at an acid or hard rock concert? The performers dress and demeanor gives clues to their source of inspiration. And the lyrics they sing can only be described as dark, extreme, and excessive. It is obvious that their music is putting too much negative energy into the body. The addicting nature of the left side reveals why this kind of music has affected so many. Time has definitely shown there is a relationship between drugs and lawlessness and hard or acid rock music.

Not only have the young been affected by this new culture, but the old as well. Interspersed throughout society we see men dressed like women and women dressed like men, rebellion replacing cooperation, hate replacing love, falsehoods replacing truth, and darkness replacing light. Many do not understand that with freedoms there are restrictions (as in the atom) on what we see, feel, and do in this physical world. They assume they have unlimited freedom. However, it is an illusion because the side effect of disobedience is bondage.

When the elements are violated the whole world suffers. Everything we do or say affects those around us in either a positive or negative way. As revealed in the following scripture, even the earth we live on is effected by our actions since it also is a living, feeling entity that is weary from the wickedness that has gone out from it. It, too, wants to rest and be cleansed from filthiness:

> "And it came to pass that Enoch looked upon the earth; and he heard a voice from the bowels thereof, saying: Wo, wo is me, the mother of men; I am pained, I am weary, because of the wickedness of my children. When shall I rest, and be cleansed from the filthiness which is gone out from me? When will my Creator sanctify me, that I may rest, and righteousness for a season abide upon my face?" (Pearl of Great Price, Moses 7:48)

We live in an electrical world that seeks balance and equilibrium. However, because too much "wanting to receive and be satisfied" negative energy has entered the system, the balance is gone. Even though we cannot see this energy, it is real and cannot be dismissed as insignificant. If one sin creates a minus (–) then multiplying a billion sins represents a lot of negative energy.

No wonder the world is in a state of tension and apprehension. With so much negative energy drawing power from the earth's positive energy, it can be felt by all. *"They know not, neither will they understand; they walk on in darkness: all the foundations of the earth are out of course."* (Bible, Psalm 82:5). God does not cause the suffering and confusion we see in the world. It comes from our refusal to obey. Our sorrows are the natural consequence of broken laws.

ONLY BY TURNING TO GOD CAN WE CURE OUR ILLS

What does this have to do with the atom? It holds the law, therefore, when its laws are broken, if we do not turn back to Him who created the law, we will suffer undesirable consequences.

Does God expect us to have order in our lives when we live in such an imperfect world? Most certainly, for this is the reason His laws exist. The elements would have remained in a state of chaos had He not organized them into a harmonious arrangement that would bless, not hinder our lives. He tells us to:

> "Organize yourselves; prepare every needful thing, and establish a house, even a house of prayer, a house of fasting, a house of faith, a house of learning, a house of glory, *a house of order*, a house of God." (Doctrine and Covenants 109:8, emphasis added)

How can this order be accomplished? By following the same basic law as set forth in the beginning atom:

> "(There)...must needs be...an opposition in all things. Wherefore, all things must needs be a compound in one; wherefore, if it should be one body it must needs remain "as dead" having no life neither death." (Book of Mormon, 2 Nephi 2:11,14)

Why would one body remain "as dead, having no life, neither death?" Since one body by itself cannot produce, the spiritual body and the physical body must work together in order to succeed. If our life is centered only on the physical world, how can the spirit part then govern and oversee the whole?

Our Spiritual Energies are Being Depleted By Sin

Even the casual observer can see that the world's spiritual energies are being depleted by sin. Society's problems are escalating because individuals are focusing mainly on the gratification of their physical needs rather than their spiritual needs. Without a spiritual reawakening, the world will remain "as dead, having no life, neither death." This of course means that things will still exist, but for what purpose will they exist? If the spiritual part that brings meaning and reason to our lives is no longer the strong force that governs the whole, who will govern our elements? Of course the physical part will. But when the flesh rules there is chaos.

At the root of the problem is a lack of reverence for God. As anyone knows who has watched television, been to the movies, or listened to the public's everyday conversation these past few years, speaking lightly of spiritual things is one of the great sins of our day. As confusion and turmoil increase in the world, there is an ever-increasing irreverence for God and His creations. Need we be reminded of the first and great commandment?

> "Thou shalt love the Lord thy God with all thy heart, with all thy might, mind and strength; and in the name of Jesus Christ thou shalt serve him." (D & C 59:5)

One might think this is oversimplifying the matter by saying that loving God will cure our ills. But is it? If we love God, we will show reverence to Him and His creations. We would never say anything to offend another person or think of harming anyone in any way. We would never be crude or ill mannered, for these unseemly behaviors would be counter to God's nature. Yet this is what we are seeing everywhere. Man curses and swears in His name and publicly discusses that which should be held sacred. He speaks in crude, vulgar tones about the sacred reproductive processes of the body and laughs and makes light of that which God has ordained. But this is no laughing matter, for by doing so we offend our Great Creator. Since we have been created in His image and have His spark of divinity within us, should we speak lightly of sacred things, this irreverence will create a minus that takes away from our spirits, instead of a plus that makes them strong. We, in essence, will "remain as dead," since nothing will be governing our elements, but the flesh. There will still be some life, but not what could be had. Life will remain meaningless unless the spirit is strengthened and oversees the whole.

The Will of the Flesh Leads to Terrible Consequences

How powerful the devil is when he can get his foot in the door and persuade us to do the will of the flesh. But it is deception, for when the flesh rules, it leads to bondage. Every beer, alcohol, and tobacco advertisement tries to convince us that the will of the flesh will bring us the "good times." But do not be deceived. All Satan's "good times" will do, is cause the physical part to draw power from the spiritual part, thus enabling him to lead us slowly down to hell where he can then rule and reign over us in his kingdom. As history attests, great nations have fallen because it's citizens were "ripened in iniquity." Their fall was inevitable because their excesses could not be sustained. People ripened in iniquity no longer fear the consequences of broken laws. They scoff at God's laws, not realizing that His laws are within them and that each act of disobedience is slowly taking its toll, not just on themselves, but on everyone. The world is again witnessing this same phenomenon.

We cannot blame God for our ills, for they are a natural result of rebellion on a wide scale. With the world's energies out of sync, all can feel it. Things are in disarray because the world is being governed by man's thinking, not God's thinking. The sacred is being maligned because man is choosing the desires of the flesh over the desires of the spirit. Remember this next scripture?

> "Men are free according to the flesh...(to)...choose eternal life, according to the will of the Holy Spirit. And not choose eternal death, according to the will of the flesh." (Book of Mormon, 2 Nephi 2:27-29)

The world is suffering because of the "will of the flesh." A homosexual, a drug user, and a pornography addict want immediate gratification; a despot wants power; a thief wants material gain; a liar wants to hide his shortcomings; and an adulterer wants fleeting physical pleasure. But it must be remembered that these big problems do not start within a vacuum. They first start with uncontrolled thoughts within the mind. We first choose to be selfish, rather than unselfish, and to let our minds wander, rather than control our seeds of discontent. We first choose to covet, rather than give thanks for what we have. We first choose unrighteous dominion, rather than serving the needs of others, and we first want it our way, rather than submit to the Father's will.

We are having problems because man is not controlling his thoughts nor is he submitting to the Father's will. But if one does not know he is a child of God, he is at a disadvantage. He will conduct himself as the beast he thinks he is descended from and his unbridled left side will become his worst enemy. But if he knows he is descended from a Heavenly Father who loves him and wants him to become like Him, he will control this left side; knowing full well that it can become his best friend. This is the key. We must know "from whence cometh man."

God is the Source of Our Blessings

When we look at the magnificent buildings, space ships, and computers man has created, we are led to say, "Look what man hath wrought; surely he is an amazing creature, for how else could such things been brought about?" How blind we are if we think these extraordinary feats came about because of man's brilliance. God has illuminated his mind to the laws, which brought about this success. It was not by man's reasoning that the world was created or that by his reasoning we enjoy the blessing of life. Our Heavenly Father and His Son are the source of all that is good and true. These marvels of science have come forth at this time because an Omnipotent Creator has guided man's puny arm and allowed them to come forth. All truth comes from God and without His continued blessings, man will fail.

The Father and the Son must surely be growing weary of our inability to comprehend who they are and what they are trying to accomplish here on earth. They have given us commandments so that we might know the parameters of the law, yet, many consider them irrelevant in this day of sophistication and continue to toss them aside. They would rather suffer the consequences of sin than recognize that God does not allow sin. Society has tried every conceivable way to negate the consequences of sin, but when we break God's laws, the affects of our disobedience will eventually spring forth to our own damnation. It is time that we, the citizens of this great nation, repent and humble ourselves. Each individual counts! This is how sin has spread so rapidly throughout the world, and by each individual it must now be contained. If we will but listen to and obey the words that God has given us, He will then infuse us with a power much greater than we now collectively hold.

GOD EXISTS AND HE IS AT THE HELM

What does this have to do with the atom? Just as within the atom, God has set the laws and boundaries in which we must operate.

There is a God in Heaven and He is at the helm! Because He is the Father of our spirits, He is solicitous of our well-being. He has not left us rudderless and without bearings in this endless sea of matter, nor floundering from lack of direction; but has been actively involved in charting our course on the unknown sea. From the beginning He has spoken to us in three different ways; through whisperings and promptings of the Spirit, through answers to individual prayers, and through His chosen prophets. This relaying of information from God to man has continued throughout history and has played a vital role in bringing strength and stability to a fallen world.

God revealed His will to holy men of old so that His children might know the set laws and boundaries in which they must operate. Their words have been recorded throughout history in the books, scrolls, and papyrus contemporary to their time. Because of the timelessness of their messages they continue to bless people's lives. To those who have listened and heeded their words, it has been their compass and sure direction—to those who have not, they have been alone, charting their own course.

Joseph Smith, the founder of The Church of Jesus Christ of Latter-day Saints, was also a prophet of God. He, too, received revelation from the Lord, just as these holy men of old. However, he was given an additional calling to restore all things: "*I am the Lord thy God, and I gave unto thee, my servant Joseph, an appointment, and restore all things.*" (D&C 132:40) Joseph's special calling was to *restore to* the earth the *authority* these earlier prophets held, and to bring back knowledge they had received that was forgotten or lost. The inspired words he restored are found in the Book of Mormon, Doctrine and Covenants, and Pearl of Great Price. They are the contemporary books of our time and are just as scriptural as the Bible of the past. Each stands as a witness of his prophetic calling.

Among the truths that Joseph Smith restored, was the knowledge that "there is an opposition in all things"—both things "to act" and things "to be acted upon," and that all things are a "compound in one." (Book of Mormon, 2 Nephi 2:11,14.) He learned of these truths while translating the Book of Mormon from ancient records in 1829. He also learned that these words had originally been revealed to a prophet named Lehi 2,500 years before, but because of wars and contentions among the people threatening their existence, the records were hidden from man's view. This is why a restoration was needed to bring them back.

Joseph also would not have known how prophetic and timely this next scripture would be, when it was revealed to him on 16 December 1833:

> "Yea, verily I say unto you, in that day when the Lord shall come, *he shall reveal all things*. Things which have passed, *and hidden things which no man knew, things of the earth, by which it was made*, and the purpose and the end thereof." (Doctrine and Covenants 101:32,33, emphasis added)

The list of evidences that God knows all things, even before they come to pass, could go on and on. I, therefore, challenge you to prayerfully read these scriptures that Joseph Smith restored so that you, too, might know that the prophecies contained therein came from God and are true and worthy of your consideration.

Combining Verbal Words With Visual Pictures

By combining the verbal words of a prophet with the visual pictures of the atom, truth is enlarged and magnified. The Pawlowski Atom gives stunning evidence that Jesus Christ, under the direction of his Father, created the world by a law of opposites that is firm and unchanging. With this new knowledge, we can better understand how he organized the elements so that heaven and earth could work together as one, and why he wants our elements to do the same. With freedom (expansion) on the one hand, yet restriction (contraction) on the other, if we are to achieve the oneness that he desires, this pattern must govern our lives.

Can there be any doubt? The atom, as well as the scriptures, are telling us that freedom cannot be had without restriction. These are not man's words, this is what these two witnesses reveal. It is not debatable, it is true. The puzzling aspects of our lives can be explained only when we realize there is a law of opposites governing our spirit and physical bodies. This opposite nature explains why some scriptural utterances seem contradictory. With two types of bodies to be considered, each must be dealt with in a different way.

Just as positive and negative energies are different, yet compatible, our spiritual and physical bodies are different, yet compatible. This compatibility comes from working under the same system. Since one has the freedom "to act," and the other is restricted and has "to be acted upon," the rules governing the two will be different. This is why we have commandments and why prophets and apostles are needed to guide the people in truth. Commandments inform us of what we can or cannot do in this electrically charged physical world, and prophets and apostles warn us of the consequences that come when spiritual and physical energies are off balance and out of sync. With the world in confusion and disarray, we find ourselves at just such a time. To resist the excesses of the physical world, the need for their direction is apparent.

As a final testament of prophets and their role in revealing the law, you must know that there is a prophet living in our midst today who speaks for God! He is President Gordon B. Hinckley of The Church of Jesus Christ of Latter-day Saints. His inspired words have blessed the lives of millions of people in these unsettling times, and if you open your hearts to his messages, they will also bless yours.

ALL TRUTHS TO BE GATHERED INTO ONE

What does this have to do with the atom? The scriptures testify that Jesus Christ created the world. The physical world now testifies this is true.

Why is there confusion and uncertainty in the world? I believe it is because religion, the heavenly part and science, the earthly part have no bonding power. They are not "pulling" together as God intended. There is no cohesiveness or oneness between the two. Because they are out of round, it is causing the world to be out of round. They, of necessity, must now come together to form a united whole. And why shouldn't they? Since the elements are working together as one, shouldn't they also be doing the same? The answer is obvious. They should be and in due time they will be, for the process of unification has already begun. We learn from Ephesians 1:10 in the Bible, that in the last days, God's work among the children of men will culminate in His "gather(ing) together in one, all things in Christ, *both in heaven and on earth*."

For this gathering to be accomplished, *all things must testify of Christ*, even the truths of the physical world! But this is so foreign to our thinking. How can the physical world testify of Christ when science has convinced so many that things happen by chance, not grand design? It is true this is not the thinking of the scientific world, but as previously stated, it must happen so that Christ's mission can be fulfilled.

Since the right positive strong force governs and oversees the whole, of necessity, the truths of the religious world had to first be restored before the truths of the physical world could be gathered in. This was accomplished in 1830 when Jesus Christ brought his ancient church back to earth. He named his newly restored church, The Church of Jesus Christ of Latter-day Saints. This restoration of truths from the spiritual realm (heaven) has been the pivotal point in this gathering together in one, all things in Christ. Its fruits will in time, unite all men into one great family of nations, each with different cultures, yet all worshipping together as one.

Science now needs a unifying power so that error will be cast away and only truth will prevail. Since Jesus Christ was the creator of the world and atoms and elements the material that he used to accomplish this awesome task, anything that enlarges or enhances our knowledge of this process, will be part of this gathering and unifying of truth.

Grand Unification: The Uniting of Science to True Religion

As we approach the Second Coming of Christ and the last days of the earth in its present state, we see an interesting parallel at work. While God has been working to unify His children into one great family, science has been working at unifying its diverse theories into one "Grand Unifying Theory." From Einstein in the past to Stephen Hawking in the present, scientists believe a grand unifying theory exists that will pull its diverse theories into one all encompassing theory that will explain the universe. Even though it has eluded them in the past, they believe it will yet be found. The following quote tells of this search:

> Confined to his wheelchair, unable to speak, physicist Stephen Hawking seeks the Grand Unification theory that will explain the universe.
>
> ...Specifically, Hawking is one of a number of scientists searching for the magnificently named Grand Unification, *a theory linking the two greatest intellectual achievements of the 20th century, relativity and quantum mechanics*. The former deals with the large-scale structure of the universe, as determined essentially by gravity; the latter is concerned with the forces that operate at the atomic scale and below. Reconciling these two theories, a goal that eluded even Einstein, might hold the key to understanding how the universe came into being. A grand unification theory would deserve not just a Nobel Prize but the last Nobel Prize; as Hawking puts it, "There would still be lots to do (in physics), but it would be like mountaineering after Everest." (*Newsweek*, "Reading God's Mind," June 13, 1988, pg. 56, emphasis added)

I believe the Pawlowski Atom will yet lead science to this long, sought Grand Unifying Theory, for embodied within its structure are the fundamental truths that will unlock the door. It is conceivable that the knowledge it imparts will be the means of uniting science with true religion! Now this might seem an impossible task, but once you see how visualizing the atom opens the mind to thoughts that might never have been considered, you will understand why its biggest hurdle has been overcome. *It changes one's thinking.*

One would then ask, "Could religion and science possibly be a separate part of the same system and could God possibly be at the helm?" And with their opposite natures more clearly defined, one could then visualize how the freedom of the right energy field is why religion (+), the great motivator that moves people "to act and do good works" is found within its boundaries. And the restriction of the left energy field could then more logically be attached to science, since it is the study of physical elements that must be "acted upon" (–) and the receiver of the motivation coming from the right energy field.

See how they work hand in hand? It is a win-win situation for both since you cannot have one without the other. By putting religion and science side by side, the world will eventually know that God knew what He was doing all along. Should such a paradox be true, I believe they will ultimately stand together as witnesses of the mighty and wondrous works that have been wrought by the Father and the Son!

The Law of Opposites is the Law of Balance and Equilibrium

The *law of opposites*, as shown in the Pawlowski Atom, is in reality the *law of balance and equilibrium*. It is the law embodied within our elements. It is the law that brings wholeness to our being. Because it is the law of cause and effect, we of necessity must obey this law or suffer its undesirable consequences. Obviously, everything depends upon the choices we make here in mortality. Peace will come into our lives if we learn to obey God's laws, whereas unrest and confusion if we refuse. Even though we live in a world that has lost its balance and equilibrium, if we individually work at bringing our own spirit and physical bodies into balance and harmony, there is hope that this oneness will spread until eventually the world will be doing the same.

Since the Pawlowski Atom is a physical witness, could it be women's part that balances out and completes the whole? For surely the order of the heavens requires that man be the right, spiritual witness, and woman, the left, physical witness.

Grand Unification! What a perfect name for truth that unites two parts. Positive to negative—spirit body to physical body—men to women—cosmos to subatomic—relativity to quantum—religion to science—heaven to earth; all uniting with strong bonds of power that pulls truth into a completed whole. Only a catalyst like the Pawlowski Atom could bring these opposites into a oneness of purpose because only the Pawlowski Atom reveals the exactness of law.

Yes, it is Grand Unification! By having the spiritual and physical world unite, strength and unity will be brought to a divisive, fragmented world. This unification will yet testify that Jesus Christ, as the central figure of this world, as well as its Creator, is governing the wonders of the heavens and the bounties of the earth! My prayer is that all who desire it, will be prepared to meet him when he comes again.